Canadian Mathematical Society
Société mathématique du Canada

Springer Science+Business Media, LLC

CMS Books in Mathematics

Ouvrages de mathématiques de la SMC

1 HERMAN/KUČERA/ŠIMŠA Equations and Inequalities

2 ARNOLD Abelian Groups and Representations of Finite Partially Ordered Sets

3 BORWEIN/LEWIS Convex Analysis and Nonlinear Optimization

4 LEVIN/LUBINSKY Orthogonal Polynomials for Exponential Weights

5 KANE Reflection Groups and Invariant Theory

6 PHILLIPS Two Millennia of Mathematics

7 DEUTSCH Best Approximation in Inner Product Spaces

8 FABIAN ET AL. Functional Analysis and Infinite-Dimensional Geometry

9 KŘÍŽEK/LUCA/SOMER 17 Lectures on Fermat Numbers

10 BORWEIN Computational Excursions in Analysis and Number Theory

11 REED/SALES Recent Advances in Algorithms and Combinatorics

12 HERMAN/KUČERA/ŠIMŠA Counting and Configurations

13 NAZARETH Differentiable Optimization and Equation Solving

14 PHILLIPS Interpolation and Approximation by Polynomials

John Lawrence Nazareth

Differentiable Optimization and Equation Solving

A Treatise on Algorithmic Science and the Karmarkar Revolution

With 14 Illustrations

Springer

John Lawrence Nazareth
Professor
Department of Pure and Applied Mathematics
Washington State University
Pullman, WA 99164-3113
USA
nazareth@amath.washington.edu

and
Affiliate Professor
Department of Applied Mathematics
University of Washington
Seattle, WA 98195
USA

Mathematics Subject Classification (2000): 34G20, 65K05, 68Wxx, 90C05, 90C30

Library of Congress Cataloging-in-Publication Data
Nazareth, J.L. (John Lawrence)
Differentiable optimization and equation solving : a treatise on algorithmic science and the Karmarkar revolution / John Lawrence Nazareth.
p. cm.—(CMS books in mathematics ; 13)
Includes bibliographical references and index.
ISBN 978-1-4419-3061-3 ISBN 978-0-387-21788-8 (eBook)
DOI 10.1007/978-0-387-21788-8

1. Mathematical optimization. 2. Programming (Mathematics) 3. Algorithms. I. Title. II. Series.
QA402.5 .N375 2003
519.3—dc21 2002030238

ISBN 978-1-4419-3061-3 Printed on acid-free paper.

Originally published by Springer-Verlag New York, Inc. in 2003
Softcover reprint of the hardcover 1st edition 2003

9 8 7 6 5 4 3 2 1 SPIN 10891209

Typesetting: Pages created by the author using a Springer TeX macro package

www.springer-ny.com

To my parents,
John Maximian Nazareth (in memoriam)
and
Monica Freitas-Nazareth

Preface

In 1984, N. Karmarkar published a seminal paper on algorithmic linear programming. During the subsequent decade, it stimulated a huge outpouring of new algorithmic results by researchers worldwide in many areas of mathematical programming and numerical computation.

This monograph gives an overview of a resulting dramatic reorganization that has occurred in one of these areas: algorithmic differentiable optimization and equation-solving, or, more simply, *algorithmic differentiable programming.* A new portrait of this field has emerged in recent years, which is summarized in the opening chapter and then considered, in more detail, in subsequent chapters.

The primary aim of this monograph is to provide a unified perspective and readable commentary on the above subject, with emphasis being placed on the problems that form its foundation, namely, unconstrained minimization, solving nonlinear equations, the special case of unidimensional programming, and linear programming. The work discussed here derives, in the main, from the author's research in these areas during the post-Karmarkar period, and the focus of attention throughout the monograph will be on the following topics:

- the *formulation of root-level algorithms* with a view to developing an understanding of their interrelationships and underlying mathematical substructure;
- the *illustration of their performance* via basic computational experiments on a mix of contrived and/or realistic test problems;

- a consideration of larger issues that center on the emerging discipline of *algorithmic science.*

For a detailed overview, see Chapter 1 and, in particular, Section 1.3.

The convergence-and-complexity analysis of optimization and equation-solving algorithms considered here is important from a mathematical standpoint. However, this facet of the subject does not lie within our main sphere of interest and it is treated only very briefly. The interested reader can find an in-depth and up-to-date background discussion in several recently published books, for example, Bertsekas [1999], Blum, Cucker, Shub, and Smale [1998], and Kelley [1999]. For additional background perspective on convergence analysis, see also Powell [1991].

The reader is assumed to be familiar with advanced calculus; numerical analysis, in particular, numerical linear algebra; the theory and algorithms of linear and nonlinear programming; and the fundamentals of computer science, in particular, computer programming and the basic models of computation and complexity theory. Thus, this monograph is intended for researchers in optimization and advanced graduate students. But others as well will find the ideas to be of interest.

The book can be used for self-study, as a research seminar text (accompanied by selected readings from the cited literature), or as a supplement to a comprehensive graduate-course textbook on optimization and equation-solving, for example, Bertsekas [1999]. A variety of open avenues of algorithmic research are highlighted in many of the chapters, which can provide the basis for research projects in a seminar or graduate course. And if the typical reader comes away with a refreshing, new perspective on algorithmic differentiable programming and an appreciation of the joys of algorithmic science[1] in general, then my objectives in writing this monograph will more than have been fulfilled.

I feel very fortunate to have been able to participate in the post-Karmarkar algorithmic revolution. It was an extremely interesting and exciting time and afforded the opportunity to interact with many researchers who actively contributed to it. I have also been very fortunate to have known some of the great pioneers in the field of algorithmic differentiable optimization and equation-solving, who discovered its fundamental algorithms and have been a source of inspiration to all that followed them. I thank, in particular, three *algorithmic scientists* par excellence, who exemplify the field: Emeritus Professors George Dantzig (simplex; decomposition), Bill Davidon (variable-metric; collinear-scaling) and Al Goldstein (gradient-projection; cutting-plane).

---

[1]Webster's dictionary defines the word "treatise" (on algorithmic science), which is used in the subtitle of the monograph, as follows: "a systematic exposition or argument in writing, including a methodical discussion of the facts and principles involved and the conclusions reached."

I thank the Mathematical Sciences Research Institute (MSRI), Berkeley, California, and, in particular, Professors Michael Overton, Jim Renegar, and Mike Shub, for the invitation to participate as a member of the fall 1998 Program in Optimization. This writing was begun there with support from MSRI.

Some of the work described in Chapter 3 was done jointly with Professor Liqun Qi (University of New South Wales, Sydney), and the algorithm in Chapter 6 was developed jointly with Professor Paul Tseng (University of Washington). I thank them both for the interesting collaboration.

I'm very grateful to the University Relations Committee of the Boeing Company, Seattle, Washington, and especially to Roger Grimes, for providing seed funding for an algorithmic science feasibility study (1999–2000). Its underlying idea has roots that can be traced all the way back to my graduate studies in the Departments of Computer Science and IE&OR at Berkeley (1968–1973), and I thank my research advisors Professors Beresford Parlett and Stuart Dreyfus for their wise counsel during and after these formative years.

I gratefully acknowledge my home Department of Pure and Applied Mathematics, Washington State University, Pullman, and my affiliate Department of Applied Mathematics, University of Washington, Seattle, and, in particular, Professors Mike Kallaher, Bob O'Malley, Terry Rockafellar, and K.K. Tung, for providing nurturing and collegiate research environments. A sabbatical leave (2000–2001), which was spent at this interface, gave me the opportunity to do much of the writing and bring the work to completion. I also thank Professor Jonathan Borwein of Simon Fraser University and Mark Spencer, Margaret Mitchell and Jenny Wolkowicki of Springer-Verlag, New York, for facilitating the book's publication.

Finally, a special thanks to my wife, Abbey, for her encouragement and helpful advice on this project, and, more importantly, for the joy that she brings to life.

John Lawrence Nazareth
Bainbridge Island, 2002

Contents

V Algorithmic Science 165

Comments on Notation

Euclidean space of dimension n, over the real numbers R, is denoted by R^n, where n is a positive integer. The non-negative orthant of R^n is denoted by R^n_+ and the *positive* orthant by R^n_{++}. Also $R^1 \equiv R$.

Lowercase italic letters denote scalars, e.g., $x \in R$, or scalar-valued functions, e.g., $h : R \to R$.

Lowercase boldface letters denote vectors in R^n, e.g., $\mathbf{x} \in R^n$, or vector-valued mappings, e.g., $\mathbf{h} : R^n \to R^m$, where m is a positive integer. Lowercase italic letters with subscripts denote components of vectors, e.g., $x_i \in R$, $1 \leq i \leq n$, or components of vector-valued mappings, e.g., $h_i : R^n \to R$, $1 \leq i \leq m$.

Uppercase boldface letters denote matrices, e.g., $\mathbf{B} \in R^m \times R^n$, and lowercase double-subscripted italic letters denote their components, e.g., $b_{ij} \in R$. When a matrix is symmetric, we try to use a symmetric letter whenever possible, e.g., $\mathbf{A} \in R^n \times R^n$. When it is necessary to emphasize that a matrix is also positive definite, we attach a '+' superscript, e.g., $\mathbf{A}^+$.

Lowercase Greek letters denote scalars or vectors as determined by context. Uppercase Greek letters generally denote sets.

Iterates of an algorithm in R^n are identified in one of two ways:

- a subscript is attached, e.g., $\mathbf{x}_k$ or $\mathbf{B}_k$, where $k = 0, 1, 2, \ldots$.
- a superscript in parentheses is attached, e.g., $\mathbf{x}^{(k)}$, $k = 0, 1, 2, \ldots$.

The latter notation is employed for iterates of an algorithm in Chapters 7, 8, 10, 11, and 12 of Parts III and IV, which cover linear programming, because *components of iterates*, e.g., $x_i^{(k)}$ must be identified in this setting. The use of parentheses serves to distinguish $x_i^{(k)}$ from x_i^k, where the latter quantity denotes the kth power of component i of the vector $\mathbf{x}$. This alternative notation for iterates is also used in Section 3.2, which develops the foundation for the linear programming algorithms discussed in Part IV.

Part I

Foundations

1
The Karmarkar Revolution

Optimization problems seek points that maximize or minimize stated objective functions over feasible regions that are defined herein by given sets of equality and inequality constraints. Equation-solving problems seek points that simultaneously satisfy given sets of equality constraints. We restrict attention to problems where the underlying spaces are *real* and *finite-dimensional* and the functions defining objectives and constraints are *differentiable*. Problems of this type arise in all areas of science and engineering, and our focus in this monograph is on the *unified study of algorithms*[1] for solving them.

1.1 Classical Portrait of the Field

The genealogy of algorithmic optimization and equation-solving can be traced to the works of venerated mathematicians—Cauchy, Euler, Fourier, Gauss, Kantorovich, Lagrange, Newton, Poincaré, and others. But it was only after a very long gestation period that the subject was truly born in the mid nineteen forties with Dantzig's discovery of the wide-ranging practical applicability of the linear programming model—the flagship of the

---

[1]Algorithmic techniques for solving differentiable problems over R^n are also important because they provide a foundation for solving more general classes of optimization problems defined over finite- or infinite-dimensional spaces, for example, problems of nondifferentiable programming, stochastic programming, semidefinite programming, and optimal control.

field—and his invention of its main solution engine, the simplex method. Other areas of mathematical programming developed rapidly in what may fittingly be termed the *Dantzig Modeling-and-Algorithmic Revolution.* Its history is nicely told in the foreword of the recent book by Dantzig and Thapa [1997]. See also the linear programming classic Dantzig [1963].

The basic models of differentiable optimization and equation-solving within the classical treatment will now be itemized. In the discussion that follows, let f, f_i, F, h_i, and g_j denote smooth, nonlinear functions from $R^n \to R$:

Linear Programming Model

$$\begin{aligned} \text{LP: } \quad & \text{minimize}_{\mathbf{x} \in R^n} \quad \mathbf{c}^T \mathbf{x} \\ \text{s.t.} \quad & \mathbf{a}_i^T \mathbf{x} = b_i, \quad i = 1, \ldots, m, \\ & l_j \leq x_j \leq u_j, \quad j = 1, \ldots, n, \end{aligned} \tag{1.1}$$

where m and n are positive integers, $m \leq n$, $\mathbf{x}$ is an n-vector with components x_j, $\mathbf{c}$ and $\mathbf{a}_i$ are n-vectors, and $\mathbf{b}$ is an m-vector with components b_i. The quantities l_j and u_j are lower and upper bounds, which are also permitted to assume the values $-\infty$ and $+\infty$, respectively.

The special structure of the feasible region of the LP model, namely, a convex polytope, and the fact that its preeminent solution engine, the simplex algorithm, proceeds along a path of vertices of the feasible polytope, imparted the flavor of combinatorial or *discrete* mathematics to the subject of linear programming. As a result, the development of algorithmic optimization during the four decades following the Dantzig revolution was characterized by a *prominent watershed between linear programming and the remainder of the subject, nonlinear programming (NLP).* Many of the NLP solution algorithms are rooted in analytic techniques named for mathematicians mentioned above, and the subject retained the flavor of *continuous* mathematics.

Nonlinear programming, in its classical development, itself displayed a *secondary* but equally marked *watershed* between the following:

- problems with no constraints or only linear constraints, together with the closely allied areas of unconstrained nonlinear least squares and the solution of systems of nonlinear equations;
- problems with more general nonlinear equality and/or inequality constraints, including the special case where the functions defining the objective and constraints are convex.

More specifically, the problems on one side of this secondary watershed are as follows:

Unconstrained Minimization Model

$$\text{UM: } \quad \text{minimize}_{\mathbf{x} \in R^n} \quad f(\mathbf{x}). \tag{1.2}$$

Joined to UM in the classical treatment was the problem of finding a global minimum of a nonlinear function, which we will identify by the acronym GUM. An important special case of UM and GUM is the choice $n = 1$ or unidimensional optimization.

Nonlinear Least-Squares Model

$$\text{NLSQ:} \quad \text{minimize}_{\mathbf{x}\in R^n} \quad F(\mathbf{x}) = \sum_{i=1}^{l} [f_i(\mathbf{x})]^2, \tag{1.3}$$

where l is a positive integer. NLSQ is obviously a particular instance of the UM problem.

Nonlinear Equations Model

$$\text{NEQ:} \quad \text{Solve}_{\mathbf{x}\in R^n} \quad h_i(\mathbf{x}) = 0, \quad i = 1, \ldots, n. \tag{1.4}$$

The choice $n = 1$ is an important special case. A solution of NEQ is a *global* minimizing point of $\sum_{i=1}^{n} [h_i(\mathbf{x})]^2$, thereby providing the link to NLSQ, UM, and GUM.

The traditional treatment of the UM, NLSQ, and NEQ areas can be found, for example, in Dennis and Schnabel [1983].

When *linear* equality or inequality constraints are present, the algorithms for UM, NLSQ, and GUM can be extended to operate within appropriately defined affine subspaces, and the resulting subject remains closely allied. The problem is as follows:

Linear Convex-Constrained Programming Model

$$\begin{aligned} &\text{LCCP:} \quad \text{minimize}_{\mathbf{x}\in R^n} \quad f(\mathbf{x}) \\ &\text{s.t.} \quad \mathbf{a}_i^T\mathbf{x} = b_i, \quad i = 1, \ldots, m, \\ &\quad l_j \leq x_j \leq u_j, \quad j = 1, \ldots, n, \end{aligned} \tag{1.5}$$

where the associated quantities are defined as in LP above. An important special case arises when the objective function is a convex or nonconvex quadratic function, a subject known as quadratic programming (QP).

On the *other side* of the secondary watershed within nonlinear programming that was mentioned above, we have the following problem areas:

Nonlinear Equality-Constrained Programming Model

$$\begin{aligned} &\text{NECP:} \quad \text{minimize}_{\mathbf{x}\in R^n} \quad f(\mathbf{x}) \\ &\text{s.t.} \quad h_i(\mathbf{x}) = 0, \quad i = 1, \ldots, m. \end{aligned} \tag{1.6}$$

Convex-Constrained Programming Model

$$\begin{aligned} &\text{CCP:} \quad \text{minimize}_{\mathbf{x}\in R^n} \quad f(\mathbf{x}) \\ &\text{s.t.} \quad \mathbf{a}_i^T\mathbf{x} = b_i, \quad i = 1, \ldots, m, \\ &\quad g_j^c(\mathbf{x}) \leq 0, \quad j = 1, \ldots, \bar{n}, \end{aligned} \tag{1.7}$$

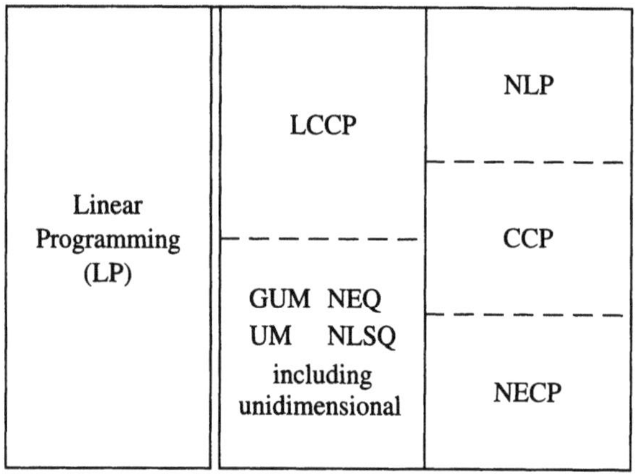

FIGURE 1.1 Classical differentiable optimization and equation-solving.

where $\bar{n}$ is an integer and $g_j^c : R^n \to R$ are smooth, convex functions.

General Nonlinear Programming Model

$$\begin{aligned} \text{NLP:} \quad & \text{minimize}_{\mathbf{x}\in R^n} \quad f(\mathbf{x}) \\ \text{s.t.} \quad & h_i(\mathbf{x}) = 0, \quad i = 1, \ldots, m, \\ & g_j(\mathbf{x}) \le 0, \quad j = 1, \ldots, \bar{n}. \end{aligned} \tag{1.8}$$

Thus, using the acronyms in (1.1)–(1.8), we can summarize the organizational picture of the subject that emerged in the four decades following the Dantzig modeling-and-algorithmic revolution by Figure 1.1.

1.2 Modern Portrait of the Field

Following a decade of intense research by many investigators worldwide, motivated by Karmarkar's 1984 article, a new portrait has emerged that has left virtually none of the foregoing problem areas untouched. Again using the acronyms defined in (1.1)–(1.8), this new portrait is summarized in Figure 1.2, and its main features are as follows:

- **UM**, NLSQ, LCCP, CCP: Algorithms for UM seek a *local* minimum, i.e., a point of local convexity of an arbitrary, smooth objective function $f(\mathbf{x})$ over the convex region R^n. The NLSQ problem is a special case of UM. The LCCP and CCP problems are natural extensions where R^n is replaced by a convex polytope and a more general convex region, respectively.
- **NEQ**, GUM, NECP, NLP: Points that satisfy a set of nonlinear equations (equality constraints) form a set that is *not* necessarily convex.

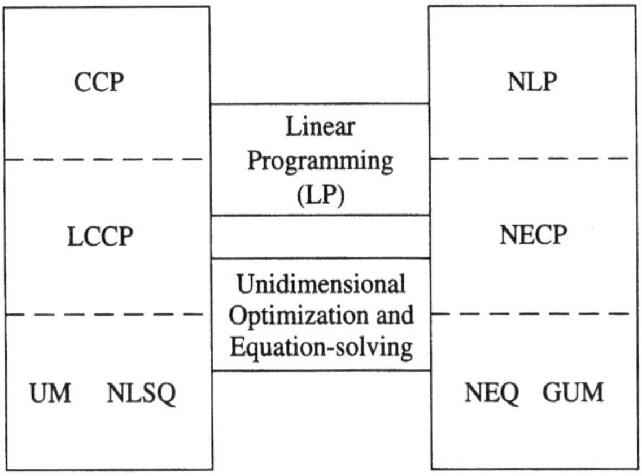

"The great watershed in optimization isn't between linearity and nonlinearity but convexity and nonconvexity" (Rockafellar [1993]).

FIGURE 1.2 Modern differentiable optimization and equation-solving.

Thus, the NEQ problem belongs naturally on the nonconvex side of the watershed. The GUM problem must take the nonconvexity of the objective function into consideration, and it also belongs naturally here. When the number of equality constraints in NEQ is fewer than the number of variables and an objective function is optimized over the resulting feasible region, one obtains the NECP problem. When inequality constraints are introduced, one obtains the NLP problem in full generality.

- **1-D:** Unidimensional optimization and equation-solving bridge the watershed depicted in Figure 1.2. Algorithms for solving UM and NEQ for the special case $n = 1$ are much more closely interrelated than their multidimensional counterparts. In the classical treatment, false lessons were learned from the 1-D case, which contributed to the incorrect grouping of UM and NEQ under a common umbrella, as depicted in Figure 1.1.
- **LP:** Linear programming is the other bridge that traverses the convexity/nonconvexity watershed. Some of its key algorithms are derived, in a natural way, by approaching the subject from the convexity side. Others are best approached from the nonconvexity side.

In this monograph, we focus on the *foundations* of our subject, namely, the **UM**, **NEQ**, **1-D**, and **LP** problems and the unified study of algorithms for solving them.[2] Because the areas of linear programming, nonlinear pro-

---

[2] Other problems in Figure 1.2 build on these foundations. For a comprehensive and up-to-date treatment that fits the overall pattern of Figure 1.2, see Bertsekas [1999].

gramming, and nonlinear equation-solving are so closely interconnected in the post-Karmarkar treatment, it is appropriate that the entire subject be called *differentiable optimization and equation-solving*, or, more compactly, *differentiable programming.*[3]

1.3 Overview of Monograph

Our discussion falls into five parts, each comprising three chapters, as follows:

Part I: In the next two chapters, we consider algorithms for the unconstrained minimization and nonlinear equation-solving problems, respectively.

In Chapter 2, we show that a single underlying technique, the Newton–Cauchy (NC) method, lies at the heart of most unconstrained minimization algorithms in current use. It combines a model in the spirit of Newton's method and a metric in the spirit of Cauchy's.

In Chapter 3, we first give examples to illustrate the well-known fact that solving nonlinear equations via nonlinear least squares is useful in a practical sense, but inherently flawed from a conceptual standpoint. Homotopy-based Euler–Newton (or predictor/corrector) techniques, neglected in the classical treatment in favor of techniques based on the sum-of-squares merit function (Figure 1.1), must be given a more prominent role at the algorithmic foundations of the subject. A natural association between nonlinear equation-solving and nonlinear equality-constrained programming (NECP) reveals a fundamental alternative, the Lagrange–NC method. It employs Lagrangian-based potentials derived from standard NECP techniques, and uses them, in conjunction with Newton–Cauchy algorithms, to attack the nonlinear equation-solving problem.

Part II, in three chapters, covers lessons that can be learned from unidimensional programming.

The first, Chapter 4, explains why algorithms for solving unconstrained minimization and nonlinear equation-solving for the special case $n = 1$ are much more closely interrelated than their multidimensional counterparts.

Next, the key role of the unidimensional line search in the fortune of the Fletcher–Reeves nonlinear conjugate-gradient algorithm is the topic of Chapter 5.

A third unidimensional lesson is considered in Chapter 6. Here, a link is established between classical golden-section search and the generic Nelder–

[3]These alternative names also provide a nice counterpoint to the widely studied area of nondifferentiable (or nonsmooth) optimization and equation-solving.

Mead (NM) direct-search algorithm restricted to $n = 1$, in order to obtain a conceptually more satisfactory approach to the multidimensional NM algorithm.

Part III addresses the linear programming problem from the perspective of the left half of the watershed of Figure 1.2.

Both the simplex algorithm of Dantzig and the affine-scaling interior algorithm of Dikin can be conveniently approached from the convexity side. The simplex method is a highly specialized form of active-set method for linearly constrained convex programming. And the NC model/metric approach to unconstrained minimization of Chapter 2, restricted to appropriate diagonal matrices, carries across to linear programming and yields Dikin's algorithm. These key algorithms along with dual and primal–dual variants are the focus of Chapters 7 and 8. The complexity of linear programming algorithms discussed in these two chapters is *not* known to be polynomial.

Diagonal metrics for *nonlinear* unconstrained minimization problems, under the rubric quasi-Cauchy (QC), are also considered in this part of the monograph, in its concluding Chapter 9.

Part IV again addresses the linear programming problem, now approached from the right half of the watershed of Figure 1.2.

The homotopy-based Euler–Newton method of Chapter 3, applied to the Karush–Kuhn–Tucker optimality equations of a linear program, yields a variety of basic path-following interior algorithms discussed in Chapter 10.

Extracting the fundamental connection with the classical logarithmic-barrier approach to nonlinear programming is the topic of Chapter 11.

Barrier functions, in turn, motivate potential functions—the centerpiece of Karmarkar's algorithm—as discussed in Chapter 12.

Linear programming algorithms discussed in this part of the monograph usually exhibit polynomial complexity.

Part V: Parts I–IV of the monograph provide a vehicle for considering larger issues, the topic of the last three chapters.

Within the specific setting of the unconstrained minimization problem, Chapter 13 considers basic algorithmic principles on which variable-metric and related algorithms are premised. Implications, in general, for other families of algorithms are also discussed.

Chapter 14 develops a new paradigm for optimization and equation-solving based on the fundamental Darwinian ideas of population-thinking and variation.

Finally, Chapter 15 concludes the monograph with a philosophy and vision for the emerging discipline of algorithmic science.

1.4 Notes

Section 1.2: Figure 1.2 and the post-Karmarkar restructuring are based on Nazareth [2000].

2

The Newton–Cauchy Method

Unconstrained optimization algorithms seek a local minimum of a nonlinear function $f : R^n \rightarrow R$, where f is smooth. This basic problem and its associated algorithms are an important part of the foundation of differentiable programming, as portrayed in Figure 1.2.

The classical techniques named for Newton and Cauchy view the unconstrained minimization problem from complementary perspectives, model-based and metric-based, respectively. They provide a coherent framework that relates the basic algorithms of the subject to one another and reveals hitherto unexplored avenues for further development. We will see that a single method, which we call the Newton–Cauchy (NC) method, subsumes most unconstrained minimization algorithms in current use and suggests some interesting new ones.

2.1 Introduction

Consider the unconstrained minimization problem for the case $n = 2$. A simple and well-known example is Rosenbrock's function,

$$f(\mathbf{x}) = (1 - x_1)^2 + 100(x_2 - x_1^2)^2, \tag{2.1}$$

which defines a steep curving parabolic valley with a minimum at the point $(1, 1)$. When this function is used as a test problem, the standard starting point is $(-1.2, 1)$. This forces an algorithm to navigate the steep valley and, in particular, the challenging topography in the neighborhood of the origin.

An oft-used algorithmic metaphor for the case $n = 2$ is that of a marble rolling downhill to a minimizing point on a mountainous, Rosenbrock-like landscape. However, this metaphor is misleading. On its downward path, a marble samples the landscape for height, slope, and curvature, continuously and at no expense. Thus, the metaphor ignores a central tenet of algorithmic optimization, namely, that the acquisition of information at any point $\mathbf{x}$ incurs *a significant, nonzero cost.* In practical applications, or in simulated settings using test problems, this cost often outweighs *all* other costs associated with manipulating information within the unconstrained minimization algorithm; see also Section 2.1.1 below.

The principal challenge of algorithmic optimization is to find a solution with as few requests for information as possible. Thus, a much better metaphor, again for the case $n = 2$, is that of a small boat floating on an *opaque* lake that entirely covers the landscape. At the current location of the boat, the depth of the lake, the slope, and perhaps even the curvature can be obtained, each at a known cost. The boatman's task is to use the information that he chooses to gather to move the boat across the surface to a new and better location, where the landscape can be sampled again. The process is repeated, with the aim of finding the deepest point[1] of the lake and using as few depth/slope/curvature samples as possible.

2.1.1 *Informational Complexity Assumptions*

A nonlinear unconstrained optimization algorithm is often assumed to operate under the following two basic assumptions:

1. *The cost of function/gradient/Hessian information*—henceforth the symbol $f/g/H$ denotes one or more of these quantities—*overshadows the cost of all other computation within an algorithm.* Even when a simple and very inexpensive-to-compute test function is used to evaluate performance of an algorithm, a common measure of performance is the number of calls to the $f/g/H$ evaluation routine. This *simulates* a practical situation where an objective function resembles the test problem and the computational cost of obtaining the $f/g/H$ information is very high. For example, the function contours of a practical problem could be similar to the contours of Rosenbrock's function, but obtaining the associated information might involve a very expensive computation, for example, the solution of a set of partial differential equations.

2. *The $f/g/H$ routine is difficult to parallelize.* For example, a Fortran routine that computes the value of f at a given iterate, say $\mathbf{x}_k$, and is written for a sequential machine may be used by an automatic

[1] In the sense of a local minimum.

differentiation package to automatically generate Fortran routines for computing the gradient and Hessian. Neither the original routine nor the generated routines may lend themselves to parallelization.

The goal of an unconstrained minimization algorithm is to find an acceptable solution to a given problem by making as few calls to the $f/g/H$ routine as possible. When the number of $f/g/H$ calls is roughly the same for each iteration, then this goal is equivalent to keeping the iteration count of the algorithm as low as possible.

The following two items concerning problem difficulty are also worth noting explicitly:

- *Dimensionality*: A low-dimensional problem is *not* necessarily easier to solve than a problem of high dimension. For example, minimizing a quadratic function in a hundred variables is trivial under the assumption that the inversion of a 100×100 matrix is not an expensive operation. In comparison, a two-dimensional Rosenbrock-like function as considered in the first assumption above, whose $f/g/H$ information at any point is expensive to compute, is a much more challenging problem.
- *Constraints*: The inclusion of constraints does *not* necessarily increase the difficulty of solving a problem. For example, if one seeks to minimize Rosenbrock's problem (2.1) within a small square centered at the point $(1, 1)$, with sides of length say 0.5, then the task is facilitated, because most of the steep curving valley, in particular, the region around the origin, is excluded. Within the square feasible region, the contours of the function resemble those of a convex quadratic function. In other words, constraints may complicate the optimization *algorithm*, but they could, in fact, ease the optimization problem-solving *task*.

2.2 Model-Based Perspective

Model-based methods approximate f at a current iterate $\mathbf{x}_k$ via a local approximating model or direction-finding[2] problem (DfP), which is then used to obtain an improving point.

2.2.1 Newton

In Newton's method, the DfP is as follows:

$$\text{minimize } \mathbf{g}_k^T(\mathbf{x} - \mathbf{x}_k) + \tfrac{1}{2}(\mathbf{x} - \mathbf{x}_k)^T \mathbf{H}_k (\mathbf{x} - \mathbf{x}_k)$$

[2]We use DfP to avoid confusion with the acronym for the well-known DFP variable-metric update.

$$\text{s.t.} \quad \|\mathbf{x} - \mathbf{x}_k\|_{\mathbf{D}^+} \le \delta_k, \tag{2.2}$$

where $\mathbf{g}_k$ denotes the gradient vector of f at $\mathbf{x}_k$, and $\mathbf{H}_k$ denotes its Hessian matrix, i.e., the $n \times n$ matrix of second partial derivatives $\partial^2 f/\partial x_i \partial x_j$ at $\mathbf{x}_k$. (Note that this matrix may be indefinite.) The points $\mathbf{x}$ that satisfy the quadratic constraint form the *trust region.* The quantity $\|.\|_{\mathbf{D}^+}$ denotes a vector norm defined by a positive definite symmetric matrix $\mathbf{D}^+$ that determines the scaling of variables, i.e., $\|\mathbf{z}\|_{\mathbf{D}^+} = (\mathbf{z}^T\mathbf{D}^+\mathbf{z})^{1/2}$ for any vector $\mathbf{z}$. A common choice is the Euclidean norm, where $\mathbf{D}^+$ is set to the identity matrix. The quantity δ_k is an adaptively updated parameter that defines the size of the trust region.

It can be shown that a feasible point $\mathbf{x}_*$ is the global solution of (2.2) if and only if there is a Lagrange multiplier $\lambda_* \ge 0$ such that

$$\begin{aligned} (\mathbf{H}_k + \lambda_*\mathbf{D}^+)(\mathbf{x}_* - \mathbf{x}_k) &= -\mathbf{g}_k, \\ \lambda_*(\|\mathbf{x}_* - \mathbf{x}_k\|_{\mathbf{D}^+} - \delta_k) &= 0, \end{aligned} \tag{2.3}$$

with $(\mathbf{H}_k + \lambda_*\mathbf{D}^+)$ positive semidefinite. For details of a proof, see, for example, Nazareth [1994b, p. 42-44].

A trivial case occurs when $\mathbf{x}_*$ is in the interior of the trust region, i.e., the trust-region constraint is slack, so that the DfP is essentially unconstrained and $\lambda_* = 0$. Otherwise, the constraint holds as an equality at $\mathbf{x}_*$, and henceforth, for convenience of discussion, let us assume also that the matrix $(\mathbf{H}_k+\lambda_*\mathbf{D}^+)$ is positive definite.[3] Then the optimal mutiplier is the solution of the following *one-dimensional nonlinear equation* in the variable $\lambda \ge 0$, which is derived directly from (2.3), namely,

$$\|w(\lambda)\|_{\mathbf{D}^+} = \delta_k, \qquad w(\lambda) = -(\mathbf{H}_k + \lambda\mathbf{D}^+)^{-1}\mathbf{g}_k.$$

Also, the vector $\mathbf{x}_* - \mathbf{x}_k$ is a *direction of descent* at the point $\mathbf{x}_k$. A variety of strategies can be devised for defining the new current iterate $\mathbf{x}_{k+1}$.

A pure trust-region (TR) strategy evaluates the function at $\mathbf{x}_*$. If it is not suitably improving, then the current iterate is not updated. Instead, δ_k is reduced in order to shrink the trust region and the procedure is repeated. If $\mathbf{x}_*$ is improving, then $\mathbf{x}_{k+1}$ is set to $\mathbf{x}_*$, and δ_k is suitably increased or decreased, based on a comparison of function reduction predicted by the model against actual objective function reduction.

Alternatively, the foregoing strategy can be augmented by a line search[4] along the direction of descent $\mathbf{d}_k = \mathbf{x}_* - \mathbf{x}_k$ to find an improving point, and again δ_k is revised. This *TR/LS strategy* has not been explored, in detail, to date.

---

[3]An infrequent so-called hard case that must be addressed in any practical and comprehensive algorithmic treatment arises when the matrix is only positive semidefinite. This requires a deeper analysis and a refinement of the techniques of this section. For details, see Moré [1983].

[4]For an example of a line search routine, see Chapter 5.

See also Rendl and Wolkowicz [1997] for strategies that explicitly use the dual of (2.2).

2.2.2 Quasi-Newton

When $\mathbf{H}_k$ is unavailable or too expensive to compute, it can be approximated by an $n \times n$ symmetric matrix, say, $\mathbf{M}_k$, which is used in the foregoing model-based approach (2.2) in place of $\mathbf{H}_k$. This approximation is then revised as follows. Suppose the next iterate is $\mathbf{x}_{k+1}$ and the corresponding gradient vector is $\mathbf{g}_{k+1}$, and define $\mathbf{s}_k = \mathbf{x}_{k+1} - \mathbf{x}_k$ and $\mathbf{y}_k = \mathbf{g}_{k+1} - \mathbf{g}_k$. A standard mean value theorem for vector-valued functions states that

$$\left[\int_0^1 \mathbf{H}(\mathbf{x}_k + \theta \mathbf{s}_k) d\theta\right] \mathbf{s}_k = \mathbf{y}_k; \tag{2.4}$$

i.e., the averaged Hessian matrix over the current step transforms the vector $\mathbf{s}_k$ into $\mathbf{y}_k$. In revising $\mathbf{M}_k$ to incorporate new information, it is natural to require that the updated matrix $\mathbf{M}_{k+1}$ have the same property, i.e., that it satisfy the so-called *quasi-Newton* or *secant* relation

$$\mathbf{M}_{k+1}\mathbf{s}_k = \mathbf{y}_k. \tag{2.5}$$

The symmetric rank-one (SR1) update makes the simplest possible modification to $\mathbf{M}_k$, adding to it a matrix $\kappa \mathbf{u}\mathbf{u}^T$, where κ is a real number and $\mathbf{u}$ is an n-vector. The unique matrix $\mathbf{M}_{k+1}$ of this form that satisfies (2.5) is as follows:

$$\mathbf{M}_{k+1} = \mathbf{M}_k + \frac{(\mathbf{y}_k - \mathbf{M}_k\mathbf{s}_k)(\mathbf{y}_k - \mathbf{M}_k\mathbf{s}_k)^T}{(\mathbf{y}_k - \mathbf{M}_k\mathbf{s}_k)^T\mathbf{s}_k}. \tag{2.6}$$

This update can be safeguarded when the denominator in the last expression is close to zero. A local approximating model analogous to (2.2) can be defined using the Hessian approximation in place of $\mathbf{H}_k$. The resulting model-based method is called the *symmetric rank-one*, or SR1, quasi-Newton method. For additional detail, see Conn, Gould, and Toint [1991].

2.2.3 Limited Memory

When storage is at a premium and it is not possible to store an $n \times n$ matrix, a *limited-memory symmetric rank-one* (L-SR1) approach uses the current step and a remembered set of prior steps and associated gradient changes, usually much fewer than n in number, in order to form a *compact representation* of the approximated Hessian. We will denote this approximation by L-$\mathbf{M}_k$. Details of the update can be found in Byrd et al. [1994]. An alternative approach, called a limited-memory affine *reduced Hessian* (RH) or *successive affine reduction* (SAR) technique, develops Hessian information in an affine subspace defined by the current gradient vector, the current step, and a set of zero, one, or more previous steps. Curvature estimates can be obtained in a Newton or a quasi-Newton sense. The associated methods are identified by the acronyms L-RH-N and L-RH-SR1 and the corresponding Hessian approximations by L-RH-$\mathbf{H}_k$ and L-RH-$\mathbf{M}_k$. The underlying updates can be patterned after analogous techniques described in Nazareth [1986a] and Leonard [1995], but they have not been fully explored to date.

2.2.4 Modified Cauchy

Finally, when Hessian approximations in (2.2) are restricted to (possibly indefinite) diagonal matrices $\mathbf{D}_k$, whose elements are obtained by finite differences or updating techniques, one obtains a simple method that has also not been fully explored to date (see also Chapter 9). We attach the name *modified Cauchy* to it for reasons that will become apparent in Section 2.3.

2.2.5 Summary

Each of the foregoing model-based methods utilizes a DfP at the current iterate $\mathbf{x}_k$ of the form

$$\begin{aligned} \text{minimize } & \mathbf{g}_k^T(\mathbf{x}-\mathbf{x}_k) + \tfrac{1}{2}(\mathbf{x}-\mathbf{x}_k)^T \mathcal{H}_k (\mathbf{x}-\mathbf{x}_k) \\ \text{s.t. } & \|\mathbf{x}-\mathbf{x}_k\|_{\mathbf{D}^+} \leq \delta_k, \end{aligned} \tag{2.7}$$

where $\mathcal{H}_k$ is one of the following: the Hessian matrix $\mathbf{H}_k$; an SR1 approximation $\mathbf{M}_k$ to the Hessian; a compact representation L-$\mathbf{M}_k$, L-RH-$\mathbf{H}_k$, or L-RH-$\mathbf{M}_k$; a diagonal matrix $\mathbf{D}_k$. (The other quantities in (2.7) were defined earlier.) This DfP is used in a TR or TR/LS strategy, as defined in Section 2.2.1, to obtain an improving point.

2.3 Metric-Based Perspective

Metric-based methods explicitly or implicitly perform a transformation of variables (or reconditioning) and employ a steepest-descent search vector in the transformed space. Use of the negative gradient (steepest-descent) direction to obtain an improving point was originally proposed by Cauchy.

Consider a change of variables $\tilde{\mathbf{x}} = \mathbf{R}\mathbf{x}$, where $\mathbf{R}$ is any $n \times n$ *nonsingular* matrix. Then $\mathbf{g}_k$, the gradient vector at the point $\mathbf{x}_k$, transforms to $\tilde{\mathbf{g}}_k = \mathbf{R}^{-T}\mathbf{g}_k$, which is easily verified by the chain rule. (Henceforth, we attach a tilde to transformed quantities, and whenever it is necessary to explicitly identify the matrix used to define the transformation, we write $\tilde{\mathbf{x}}_k[\mathbf{R}]$ or $\tilde{\mathbf{g}}_k[\mathbf{R}]$.) The steepest-descent direction at the current iterate $\tilde{\mathbf{x}}_k$ in the transformed space is $-\mathbf{R}^{-T}\mathbf{g}_k$, and the corresponding direction in the original space is $-[\mathbf{R}^T\mathbf{R}]^{-1}\mathbf{g}_k$.

2.3.1 Cauchy

Let $\mathbf{D}_k^+$ be a positive definite, *diagonal* matrix, which is either fixed or can be varied at each iteration, and let $\mathbf{D}_k^+ = (\mathbf{D}_k^{1/2})^2$ be its square-root, or Cholesky, factorization. When the above matrix $\mathbf{R}$ is defined by $\mathbf{R} = \mathbf{D}_k^{1/2}$, corresponding to a rescaling of the variables, one obtains Cauchy's method. A line search along the direction of descent $-(\mathbf{D}_k^+)^{-1}\mathbf{g}_k$ yields an improving point, where the procedure can be repeated. Techniques for developing $\mathbf{D}_k^+$ will be considered in Chapter 9; see, in particular, expressions (9.5)–(9.6).

2.3.2 Variable Metric

Consider next the case where the matrix defining the transformation of variables is an $n \times n$ matrix $\mathbf{R}_k$ that can be changed at each iteration. Suppose a line search procedure along the corresponding direction $-[\mathbf{R}_k^T\mathbf{R}_k]^{-1}\mathbf{g}_k$ yields a step to an improving point $\mathbf{x}_{k+1}$, and again define $\mathbf{s}_k = \mathbf{x}_{k+1} - \mathbf{x}_k$ and $\mathbf{y}_k = \mathbf{g}_{k+1} - \mathbf{g}_k$. How should we revise the reconditioner $\mathbf{R}_k$ to $\mathbf{R}_{k+1}$ in order to reflect new information? Ideally, the transformed function, say $\tilde{f}$, should have concentric contour lines, i.e., *in the metric defined by an "ideal reconditioner"* $\mathbf{R}_{k+1}$, the next iterate, obtained by a unit step from the transformed point $\tilde{\mathbf{x}}_{k+1}$ along the steepest-descent direction, should be *independent* of where $\tilde{\mathbf{x}}_{k+1}$ lies along $\tilde{\mathbf{s}}_k = \tilde{\mathbf{x}}_{k+1} - \tilde{\mathbf{x}}_k$. Such a reconditioner will not normally exist when f is nonquadratic, but it should at least have the aforementioned property at the two points $\tilde{\mathbf{x}}_k$ and $\tilde{\mathbf{x}}_{k+1}$. Thus, it is reasonable to require that $\mathbf{R}_{k+1}$ be chosen to satisfy

$$\tilde{\mathbf{x}}_k[\mathbf{R}_{k+1}] - \tilde{\mathbf{g}}_k[\mathbf{R}_{k+1}] = \tilde{\mathbf{x}}_{k+1}[\mathbf{R}_{k+1}] - \tilde{\mathbf{g}}_{k+1}[\mathbf{R}_{k+1}]. \tag{2.8}$$

This equation can be reexpressed in the original variables as follows:

$$\mathbf{R}_{k+1}\mathbf{s}_k = \mathbf{R}_{k+1}^{-T}\mathbf{y}_k. \tag{2.9}$$

For a matrix $\mathbf{R}_{k+1}$ satisfying (2.9) to exist, it is necessary and sufficient that $\mathbf{y}_k^T\mathbf{s}_k > 0$, which can always be ensured by a line search procedure. Since $\mathbf{R}_{k+1}^T\mathbf{R}_{k+1}\mathbf{s}_k = \mathbf{y}_k$, we see that (2.9) is equivalent to the quasi-Newton relation (2.5) when we impose the added restriction that the Hessian approximation be positive definite.

Consider the question of how to revise $\mathbf{R}_k$. The so-called BFGS update (named after the first letters of the surnames of its four codiscoverers) makes the simplest possible augmentation of $\mathbf{R}_k$ by adding to it a matrix $\mathbf{u}\mathbf{v}^T$ of rank one. Modifying $\mathbf{R}_k$ to $\mathbf{R}_{k+1}$ in this manner *effects a minimal change* in the sense that a subspace of dimension $n-1$ is transformed identically by the two matrices. The updated matrix $\mathbf{R}_{k+1}$ is required to satisfy (2.9) and is chosen as close as possible to $\mathbf{R}_k$, as measured by the Frobenius or spectral norm of the difference $(\mathbf{R}_{k+1} - \mathbf{R}_k)$. This update can be shown to be as follows:

$$\mathbf{R}_{k+1} = \mathbf{R}_k + \frac{\mathbf{R}_k\mathbf{s}_k}{\|\mathbf{R}_k\mathbf{s}_k\|}\left(\frac{\mathbf{y}_k}{(\mathbf{y}_k^T\mathbf{s}_k)^{1/2}} - \frac{\mathbf{R}_k^T\mathbf{R}_k\mathbf{s}_k}{\|\mathbf{R}_k\mathbf{s}_k\|}\right)^T, \tag{2.10}$$

where $\|\cdot\|$ denotes the Euclidean vector norm. For details, see, for example, Nazareth [1994b]. The descent search direction is defined as before by

$$\mathbf{d}_{k+1} = -[\mathbf{R}_{k+1}^T\mathbf{R}_{k+1}]^{-1}\mathbf{g}_{k+1}, \tag{2.11}$$

and a line search along it will yield an improving point. The foregoing BFGS algorithm is an outgrowth of seminal variable-metric ideas pioneered by Davidon [1959] and clarified by Fletcher and Powell [1963]. For other references, see, for example, the bibliographies in Bertsekas [1999] and Dennis and Schnabel [1983].

Let $\mathbf{M}_{k+1}^+ = \mathbf{R}_{k+1}^T\mathbf{R}_{k+1}$ and $\mathbf{W}_{k+1}^+ = [\mathbf{M}_{k+1}^+]^{-1}$. These two quantities can be updated directly as follows:

$$\mathbf{M}_{k+1}^+ = \mathbf{M}_k^+ - \frac{\mathbf{M}_k^+\mathbf{s}_k\mathbf{s}_k^T\mathbf{M}_k^+}{\mathbf{s}_k^T\mathbf{M}_k^+\mathbf{s}_k} + \frac{\mathbf{y}_k\mathbf{y}_k^T}{\mathbf{y}_k^T\mathbf{s}_k}, \tag{2.12}$$

$$\mathbf{W}_{k+1}^+ = \left(\mathbf{I} - \frac{\mathbf{s}_k\mathbf{y}_k^T}{\mathbf{y}_k^T\mathbf{s}_k}\right)\mathbf{W}_k^+\left(\mathbf{I} - \frac{\mathbf{s}_k\mathbf{y}_k^T}{\mathbf{y}_k^T\mathbf{s}_k}\right)^T + \frac{\mathbf{s}_k\mathbf{s}_k^T}{\mathbf{y}_k^T\mathbf{s}_k}. \tag{2.13}$$

The foregoing matrices $\mathbf{M}_{k+1}^+$ and $\mathbf{W}_{k+1}^+$ satisfy the quasi-Newton relations $\mathbf{M}_{k+1}^+\mathbf{s}_k = \mathbf{y}_k$ and $\mathbf{W}_{k+1}^+\mathbf{y}_k = \mathbf{s}_k$, respectively, and the resulting search direction is $\mathbf{d}_{k+1} = -[\mathbf{M}_{k+1}^+]^{-1}\mathbf{g}_{k+1} = -[\mathbf{W}_{k+1}^+]\mathbf{g}_{k+1}$.

2.3.3 Limited Memory

Nonlinear CG: When storage is at a premium, the BFGS algorithm can preserve steps and corresponding changes in gradients over a limited number of prior iterations and then define the matrix $\mathbf{M}_k^+$ or $\mathbf{W}_k^+$ *implicitly* in terms of these vectors, instead of explicitly by forming a square matrix. Consider the simplest case, where a single step and gradient change are preserved (called one-step memory). The update is then defined implicitly by (2.13) with $\mathbf{W}_k^+ = \gamma_k\mathbf{I}$, where γ_k is an Oren and Luenberger [1974] scaling constant, often chosen[5] to be $\mathbf{y}_k^T\mathbf{s}_k/\mathbf{y}_k^T\mathbf{y}_k$. Thus the search direction is defined by

$$\mathbf{d}_{k+1} = -\left[\left(\mathbf{I} - \frac{\mathbf{s}_k\mathbf{y}_k^T}{\mathbf{y}_k^T\mathbf{s}_k}\right)\gamma_k\mathbf{I}\left(\mathbf{I} - \frac{\mathbf{s}_k\mathbf{y}_k^T}{\mathbf{y}_k^T\mathbf{s}_k}\right)^T + \frac{\mathbf{s}_k\mathbf{s}_k^T}{\mathbf{y}_k^T\mathbf{s}_k}\right]\mathbf{g}_{k+1}.$$

---

[5]This particular choice satisfies a weakened form of the quasi-Newton relation discussed in Chapter 9. See expressions (9.3) and (9.4), transcribed for inverse Hessian approximations or complemented as in Section 13.2.1.

For algorithmic and implementational details, see Perry [1977], [1978] and Shanno [1978].

Under the assumption of exact line searches, i.e., $\mathbf{g}_{k+1}^T \mathbf{s}_k = 0$, it follows immediately that the search direction is parallel to the following vector:

$$\mathbf{d}_{k+1}^{cg} = -\mathbf{g}_{k+1} + \frac{\mathbf{y}_k^T \mathbf{g}_{k+1}}{\mathbf{y}_k^T \mathbf{s}_k} \mathbf{s}_k. \tag{2.14}$$

This is the search direction used in the Polak–Polyak–Ribière form of the nonlinear *conjugate gradient method*, which was pioneered by Hestenes and Stiefel [1952] and suitably adapted to nonlinear optimization by Fletcher and Reeves [1964]. It will be discussed in more detail in Chapter 5.

The relationship between the BFGS and CG algorithms and its fundamental role in formulating algorithms with variable storage (or limited memory) was developed by Nazareth [1976], [1979] and extended by Buckley [1978] and Kolda, O'Leary, and Nazareth [1998]. It provides the foundation for *all* CG-related limited-memory methods discussed in the present section.

VSCG and L-BFGS: More generally, a set of prior steps and gradient changes can be preserved and used in two main ways. The first is to define a variable preconditioner for the CG algorithm; see Nazareth [1979].[6] The second is analogous to the one-step memory approach just described with the added option of developing the limited-memory update recursively, the so-called L-BFGS algorithm; see Liu and Nocedal [1989]. Key implementation issues are addressed in Gilbert and Lemaréchal [1989]. An alternative compact representation for the L-BFGS update is given by Byrd et al. [1994]. Henceforth, let us denote the Hessian and inverse Hessian matrix approximations in L-BFGS by L-$\mathbf{M}_k^+$ and L-$\mathbf{W}_k^+$, respectively.

The VSCG and L-BFGS algorithms are CG-related, and they can be viewed as extending the practical "technology" of the CG and BFGS algorithms by exploiting particular algebraic expressions of the variable-metric updates that they employ.

L-RH-BFGS: More satisfactory from a conceptual standpoint is the limited-memory reduced-Hessian or successive affine reduction version of the BFGS algorithm. This approach is based on a new principle as follows: Develop BFGS-based curvature approximations in an *affine subspace*, normally of low dimension, defined by the most recent (current) gradient and step vectors, along with zero, one, or more prior gradient and/or step vectors. We will denote this algorithm by L-RH-BFGS and its Hessian approximation by L-RH-$\mathbf{M}_k^+$. It too can be shown to be CG-related. For

[6]The algorithm described in this article uses the SR1 update whenever it is positive definite, because it requires fewer stored vectors. But the BFGS update could equally well be substituted.

algorithmic details, see Nazareth [1986b], Leonard [1995], and references given therein.

L-P-BFGS: There is an alternative to L-BFGS that returns to the premises from which the BFGS was derived in Section 2.3.2. The BFGS update of the reconditioner (2.10) can be written in product form as follows:

$$\mathbf{R}_{k+1} = \mathbf{R}_k \mathbf{E}_k, \tag{2.15}$$

where

$$\mathbf{E}_k = \mathbf{I} + \frac{\mathbf{s}_k}{\|\mathbf{R}_k \mathbf{s}_k\|} \left(\frac{\mathbf{y}_k}{(\mathbf{y}_k^T \mathbf{s}_k)^{1/2}} - \frac{\mathbf{R}_k^T \mathbf{R}_k \mathbf{s}_k}{\|\mathbf{R}_k \mathbf{s}_k\|} \right)^T. \tag{2.16}$$

The matrix in parentheses, i.e., the identity modified by a matrix of rank one, is an example of an *elementary matrix.* Updating $\mathbf{R}_k$ by an elementary matrix effects a minimal change as discussed in Section 2.3.2.

Similarly, $\mathbf{R}_k = \mathbf{R}_{k-1}\mathbf{E}_{k-1}$, and recurring this expression[7] yields the following:

$$\mathbf{R}_k = \mathbf{R}_{k-j}\mathbf{E}_{k-j}\mathbf{E}_{k-j+1}\cdots\mathbf{E}_{k-1},$$

where j is any integer satisfying $1 \le j \le k-1$. When an elementary matrix is *dropped* from the foregoing expression it also effects a minimal change, because the resulting matrix, say $\overline{\mathbf{R}}_k$, and the matrix $\mathbf{R}_k$ leave a subspace of dimension $n-1$ invariant. Thus a limited-memory version of the BFGS update, which we denote by L-P-BFGS, can be obtained as follows:

1. Let j be a small integer, for example, $1 \le j \le 5$.
2. Let $\mathbf{R}_{k-j}$ be a fixed diagonal matrix, for example, the identity matrix.
3. Drop the elementary matrix $\mathbf{E}_{k-j}$ from the front of the list of elementary matrices to obtain $\overline{\mathbf{R}}_k$. Then update $\overline{\mathbf{R}}_k$ to $\mathbf{R}_{k+1}$ in the usual way; i.e., use (2.15)–(2.16) with $\overline{\mathbf{R}}_k$ replacing $\mathbf{R}_k$, and add a new elementary matrix $\mathbf{E}_k$ to the end of the list. Thus the number of elementary matrices defining the update remains a constant j.
4. Represent each elementary matrix *implicitly* by the two n-vectors in its definition. Perform matrix–vector products, for example, $\mathbf{R}_k\mathbf{s}_k$ with $\mathbf{R}_k$ represented as a product of elementary matrices, in the standard way that requires only the formation of inner products of vectors.

It can be shown that an L-P-BFGS algorithm based on the foregoing is CG-related; i.e., it reproduces the directions of the CG algorithm on

---

[7] It is interesting to observe the resemblance to the *product form of the basis* in the revised simplex algorithm.

a strictly convex quadratic and terminates in at most n steps when line searches are exact. We leave the proof as an exercise for the reader.

Many interesting issues arise in connection with organizing the computations efficiently, choosing appropriate diagonal matrices for $\mathbf{R}_{k-j}$, for example, Oren–Luenberger-type scaling $\mathbf{R}_{k-j} = \sqrt{\gamma_k}\,\mathbf{I}$, where γ_k is defined as in the L-BFGS algorithm above, choosing the elementary matrix that is dropped from *within* the list and not just from the front, and so on.

The L-P-BFGS approach is conceptually more satisfactory than the L-BFGS, and surprisingly, it has not been explored to date. A detailed formulation and comparison with other limited-memory approaches would be a worthwhile undertaking.

2.3.4 *Modified Newton*

If $\mathbf{H}_k$ is available and possibly *indefinite*, it can be modified to a positive definite matrix, $\mathbf{H}_k^+$ in a variety of ways; see, for example, Bertsekas [1999]. This modified matrix can be factored as $\mathbf{H}_k^+ = \mathbf{R}_k^T\mathbf{R}_k$ with $\mathbf{R}_k$ nonsingular, for example, by using a Cholesky factorization or an eigendecomposition. The factor $\mathbf{R}_k$ then defines a metric-based algorithm as above, a so-called *modified Newton* method (MN).

A limited-memory modified Newton algorithm analogous to L-RH-BFGS can also be formulated. For details, see Nazareth [1986c], [1994b]. Denote this CG-related algorithm by L-RH-MN and its Hessian approximation by L-RH-$\mathbf{H}_k^+$.

2.3.5 *Summary*

The steepest-descent direction in each of the foregoing methods is the direction $\mathbf{d}_k$ that minimizes $\mathbf{g}_k^T\mathbf{d}$ over all vectors $\mathbf{d}$ of constant length δ_k in an appropriate metric. (Typically $\delta_k = 1$.) Let $\mathbf{d} = \mathbf{x} - \mathbf{x}_k$. Then this DfP can equivalently be stated as follows:

$$\begin{aligned} &\text{minimize } \mathbf{g}_k^T(\mathbf{x} - \mathbf{x}_k) \\ &\text{s.t.} \quad \|\mathbf{x} - \mathbf{x}_k\|_{\mathcal{M}_k^+} \le \delta_k, \end{aligned} \tag{2.17}$$

where $\mathcal{M}_k^+$ is given by one of the following: the identity matrix $\mathbf{I}$; a positive definite diagonal matrix $\mathbf{D}_k^+$; the BFGS approximation $\mathbf{M}_k^+$; a compact representation L-$\mathbf{M}_k^+$, L-RH-$\mathbf{H}_k^+$ or L-RH-$\mathbf{M}_k^+$; a positive definite modification $\mathbf{H}_k^+$ of the Hessian matrix $\mathbf{H}_k$. The quantity δ_k determines the length of the initial step along the search direction $-[\mathcal{M}_k^+]^{-1}\mathbf{g}_k$, and a line search (LS) strategy yields an improving point.

Let us denote the inverse of $\mathcal{M}_k^+$ by $\mathcal{W}_k^+$. It is sometimes computationally more efficient to maintain the latter matrix, for example, within a limited-memory BFGS algorithm, and to define the search direction as $-[\mathcal{W}_k^+]\mathbf{g}_k$.

2.4 Newton–Cauchy Framework

A simple and elegant picture emerges from the development in Sections 2.2 and 2.3, which is summarized by Figure 2.1; see also Nazareth [1994a,b]. It is often convenient to straddle this "two-lane highway," so to speak, and to formulate algorithms based on a "middle-of-the-road" approach. We now describe the traditional synthesis based on positive definite, unconstrained models and a new synthesis, called the NC method, based on $\mathcal{M}_k^+$-metric trust regions.

2.4.1 Positive Definite Quadratic Models

At the current iterate $\mathbf{x}_k$, use an unconstrained DfP of the following form:

$$\text{minimize } \mathbf{g}_k^T(\mathbf{x} - \mathbf{x}_k) + \tfrac{1}{2}(\mathbf{x} - \mathbf{x}_k)^T \mathcal{M}_k^+ (\mathbf{x} - \mathbf{x}_k), \tag{2.18}$$

where $\mathcal{M}_k^+$ is one of the following: $\mathbf{I}$; $\mathbf{D}_k^+$; $\mathbf{M}_k^+$; L-$\mathbf{M}_k^+$; L-RH-$\mathbf{H}_k^+$; L-RH-$\mathbf{M}_k^+$; $\mathbf{H}_k^+$. Note that the DfP uses the options available in (2.17), and indeed, the foregoing quadratic model can be obtained from a Lagrangian relaxation of the latter expression. Often these options for $\mathcal{M}_k^+$ are derived directly in positive definite, quadratic model-based terms (in place of the metric-based derivations of Section 2.3). The search direction obtained from (2.18) is $\mathbf{d}_k = -[\mathcal{M}_k^+]^{-1}\mathbf{g}_k$, and a line search along it yields an improving iterate. A good discussion of this line of development can be found, for example, in Bertsekas [1999].

Model-Based	Metric-Based
Newton (N)	Modified-Newton (MN)
SR1	BFGS
L-SR1; L-RH-N; L-RH-SR1	L-BFGS; L-RH-MN; L-RH-BFGS
Modified-Cauchy (MC)	Cauchy (C)

FIGURE 2.1 Newton–Cauchy framework.

2.4.2 The NC Method

Substantial order can be brought to computational unconstrained nonlinear minimization by recognizing the existence of a *single underlying method*, henceforth called the Newton–Cauchy or NC method, which is based on a model of the form (2.7), but with its trust region now employing a metric corresponding to (2.17). This DfP takes the form

$$\begin{aligned} &\text{minimize } \mathbf{g}_k^T(\mathbf{x}-\mathbf{x}_k) + \tfrac{1}{2}(\mathbf{x}-\mathbf{x}_k)^T\mathcal{H}_k(\mathbf{x}-\mathbf{x}_k) \\ &\quad\text{s.t.} \quad \|\mathbf{x}-\mathbf{x}_k\|_{\mathcal{M}_k^+} \le \delta_k, \end{aligned} \tag{2.19}$$

where the matrix $\mathcal{H}_k$ is one of the following: the zero matrix $\mathbf{0}$; $\mathbf{H}_k$; $\mathbf{M}_k$; L-$\mathbf{M}_k$; L-RH-$\mathbf{H}_k$; L-RH-$\mathbf{M}_k$; $\mathbf{D}_k$. Note that the objective is permitted to be linear. The matrix $\mathcal{M}_k^+$ is one of the following: $\mathbf{I}$; $\mathbf{D}_k^+$; $\mathbf{M}_k^+$; L-$\mathbf{M}_k^+$; L-RH-$\mathbf{H}_k^+$; L-RH-$\mathbf{M}_k^+$; $\mathbf{H}_k^+$. Despite numerical drawbacks, it is sometimes computationally more convenient to maintain $\mathcal{W}_k^+$; also note that $\mathcal{H}_k = \mathcal{M}_k^+$ gives a model equivalent to $\mathcal{H}_k = \mathbf{0}$.

The underlying theory for trust region subproblems of the form (2.19) and techniques for computing the associated multiplier λ_k can be derived, for example, from Moré [1993] or Rendl and Wolkowicz [1997], with the next iterate being obtained by a TR/LS strategy as discussed in Section 2.2. In particular, a line search can ensure that $\mathbf{y}_k^T\mathbf{s}_k > 0$, which is needed whenever a variable metric is updated. Note also that (2.19) is the same as (2.17) when the objective function is linear, and the TR/LS strategy then reduces to a line-search strategy.

The NC method can be formulated into a large variety of individual NC algorithms. These include all the standard ones in current use, along with many new and potentially useful algorithms. *Each is a particular algorithmic expression of the same underlying method.* A sample of a few algorithms from among the many possibilities (combinations of $\mathcal{H}_k$ and $\mathcal{M}_k^+$) is given in Figure 2.2. The last column identifies each algorithm, and the symbol $*$ indicates that it is new.

Two particular avenues that are well worth further investigation are as follows:

- Formulation of the NC algorithm that uses the true Hessian to define the model and the BFGS update to define the metric, i.e., $\mathcal{H}_k = \mathbf{H}_k$ and $\mathcal{M}_k^+ = \mathbf{M}_k^+$, and comparison with a more standard implementation of the Newton algorithm.
- Formulation of the NC algorithm that uses the SR1 update to define the model and the BFGS update to define the variable metric, i.e., $\mathcal{H}_k = \mathbf{M}_k$ and $\mathcal{M}_k^+ = \mathbf{M}_k^+$, and comparison of its performance with the standard SR1 quasi-Newton and BFGS variable-metric algorithms. An implementation based on a model–metric combination of the two

$\mathcal{H}_k$	$\mathcal{M}_k^+$	Strategy	Algorithm
$\mathbf{H}_k$	$\mathbf{D}^+$	TR	Newton
$\mathbf{0}$	$\mathbf{D}_k^+$	LS	Cauchy
$\mathbf{M}_k$	$\mathbf{D}^+$	TR	SR1
$\mathbf{0}$	$\mathbf{M}_k^+$	LS	BFGS
$\mathbf{H}_k$	$\mathbf{M}_k^+$	TR/LS	*
$\mathbf{M}_k$	$\mathbf{M}_k^+$	TR/LS	*
L-$\mathbf{M}_k$	$\mathbf{D}^+$	TR	L-SR1
$\mathbf{0}$	L-$\mathbf{M}_k^+$	LS	L-BFGS
$\mathbf{0}$	L-RH-$\mathbf{M}_k^+$	LS	L-RH-BFGS
L-$\mathbf{M}_k$	L-$\mathbf{M}_k^+$	TR/LS	*
L-RH-$\mathbf{H}_k$	L-$\mathbf{M}_k^+$	TR/LS	*
L-RH-$\mathbf{H}_k$	L-RH-$\mathbf{M}_k^+$	TR/LS	*
L-RH-$\mathbf{M}_k$	$\mathbf{D}_k^+$	TR/LS	*

FIGURE 2.2 Examples of NC algorithms.

updates could very well prove to be more effective, in practice, than a model-based SR1 or a metric-based BFGS implementation.

We see that even in the relatively mature unconstrained minimization field there are still ample opportunities for new algorithmic contributions, and for associated convergence and rate of convergence analysis, numerical experimentation, and the development of high-quality software.

The Newton–Cauchy framework, summarized in Figure 2.1, will be extended in Chapter 9.

2.5 Notes

Sections 2.1–2.4: This material is based on Nazareth [2001a], [1994a,b]. The notation used in this chapter was proposed by Davidon, Mifflin, and Nazareth [1991].

3

Euler–Newton and Lagrange–NC Methods

We now turn to the second foundational problem of Chapter 1 and its overview Figure 1.2, namely, the nonlinear equation-solving problem (1.4). It can be stated more compactly as follows:

$$\text{Solve}_{\mathbf{x} \in R^n} \quad \mathbf{h}(\mathbf{x}) = \mathbf{0}, \tag{3.1}$$

where $\mathbf{h} : R^n \to R^n$. We seek a *real* solution, say $\mathbf{x}^*$, of this system of n smooth nonlinear equations in n unknowns, and we will assume that such a solution exists. The problem of finding *complex* solutions of the real system (3.1) is not addressed in this monograph.

3.1 Introduction

The basic Newton iteration for solving (3.1) takes the form

$$\mathbf{x}_{k+1} = \mathbf{x}_k - \mathbf{J}(\mathbf{x}_k)^{-1}\mathbf{h}(\mathbf{x}_k), \qquad k = 1, 2, \ldots, \tag{3.2}$$

where $\mathbf{J}(\mathbf{x}_k)$ denotes the Jacobian matrix of $\mathbf{h}$ at an iterate $\mathbf{x}_k$, and $\mathbf{x}_1$ is a given starting approximation. For example, when $\mathbf{J}(\mathbf{x}^*)$ is nonsingular and $\mathbf{x}_1$ is sufficiently close to $\mathbf{x}^*$, the iteration (3.2) is well defined; i.e., $\mathbf{J}(\mathbf{x}_k)$ is nonsingular, and $\mathbf{x}_k$ converges to $\mathbf{x}^*$ at a quadratic rate.

The Newton iteration, as just described, is equivalent to the so-called Gauss–Newton iteration for minimizing a nonlinear sum-of-squares function

$$F(\mathbf{x}) \equiv \tfrac{1}{2}\mathbf{h}(\mathbf{x})^T\mathbf{h}(\mathbf{x}). \tag{3.3}$$

This can be seen by reexpressing (3.2) as

$$\mathbf{x}_{k+1} = \mathbf{x}_k - [\mathbf{J}(\mathbf{x}_k)^T \mathbf{J}(\mathbf{x}_k)]^{-1} \mathbf{J}(\mathbf{x}_k)^T \mathbf{h}(\mathbf{x}_k), \tag{3.4}$$

and observing that $\mathbf{J}(\mathbf{x}_k)^T\mathbf{h}(\mathbf{x}_k)$ is the gradient vector of F at $\mathbf{x}_k$, and the matrix $\mathbf{J}(\mathbf{x}_k)^T\mathbf{J}(\mathbf{x}_k)$ is the Gauss–Newton approximation to the Hessian matrix of F at $\mathbf{x}_k$; i.e., the first term of the right-hand side of

$$\nabla^2 F(\mathbf{x}_k) = \mathbf{J}(\mathbf{x}_k)^T\mathbf{J}(\mathbf{x}_k) + \sum_{i=1}^{n} \mathbf{h}_i(\mathbf{x}_k)\nabla^2\mathbf{h}_i(\mathbf{x}_k).$$

The foregoing interpretation forms the basis of algorithms for solving (3.1) that incorporate line search or trust region techniques (suitably stabilized when $\mathbf{J}(\mathbf{x}_k)$ is close to singular) in order to ensure sum-of-squares merit-function decrease and force convergence from more distant starting points. However, they suffer from the weakness that iterates can be attracted to or cannot escape from *local* minima of F, where the gradient vector of this merit function vanishes, but not the mapping $\mathbf{h}$ and the corresponding residual vector of the system (3.1). This can be seen on even very simple examples. A unidimensional example can easily be found, but this is not the ideal choice, because a simple, globally convergent algorithm, namely bisection, is available for finding a zero of an equation in one variable. We will consider therefore two examples with $n \geq 2$:

1. A system of 2 equations in 2 unknowns $\mathbf{x} = (u, v)$ as follows:

$$\text{Solve } h_i(u,v) = 0, \quad i = 1, 2, u \in R^1, v \in R^1,$$

where

$$h_1(u,v) = \begin{cases} -3 - \cos(u) + e^u & \text{if } u \leq 0, \\ -3 + u + u^2 & \text{if } 0 < u \leq 2, \\ 5 + \sin(u-2) - 2e^{-2(u-2)} & \text{if } u > 2, \end{cases} \tag{3.5}$$

and

$$h_2(u,v) = v - 1. \tag{3.6}$$

Clearly, h_1, $h_2 \in C^1$, and the functions are bounded in value. Indeed, h_1 and h_2 are in C^2 everywhere except at points with $u = 2$. The system of equations has a unique solution near the point $(1.3, 1)$. The function h_1 oscillates and has multiple local minima in the regions $u \leq 0$ and $u \geq 2$, and therefore so does $F(u,v) = \frac{1}{2}[h_1(u,v)^2 + h_2(u,v)^2]$. When a standard equation-solving subroutine is used to solve this problem, for example, the IMSL library routine DNEQNJ derived from MINPACK-1 (Moré et al. [1980]), the routine converges to the solution near $(1.3, 1)$ when started from points close to it. But for most starting points, the routine terminates with the message, "The iteration has not made good progress. The user may try a new initial guess."

2. A more realistic example due to Watson, Billups, and Morgan [1987] is as follows:

$$h_k(\mathbf{x}) = x_k - \exp\left(\cos\left(k\sum_{i=1}^{n} x_i\right)\right), \quad k = 1, \ldots, n, \tag{3.7}$$

with $n = 10$ and starting point $(0, \ldots, 0)$.

When the above routine DNEQNJ is run from the starting point, it again terminates with an error message as in the previous example. The point of termination, which can be ascertained from printout inserted in the function/Jacobian user-supplied evaluation routine, is as follows:

$$\overline{x} \approx (1.704, 1.316, 1.748, 1.827, 1.956, 1.772, 1.256, 1.646, 2.539, 1.706), \tag{3.8}$$

where $F(\overline{x}) \approx 5.084$.

The quotation below from Dennis and Schnabel [1983, p. 152], which we have transcribed to use our notation, highlights the difficulties encountered in these two examples:

> There is one significant case when the global algorithms for nonlinear equations can fail. It is when a local minimizer of $F(\mathbf{x})$ is not a root of $\mathbf{h}(\mathbf{x})$. A global[1] minimization routine, started close to such a point, may converge to it. There is not much one can do in such a case, except report what happened and advise the user to try to restart nearer to a root of $\mathbf{h}(\mathbf{x})$ if possible.

The sum-of-squares, or NLSQ, merit function, when used to globalize Newton's method for equation-solving, is thus *inherently flawed.* Nevertheless, techniques based on it remain very useful from a practical standpoint, and newer variants continue to appear, for example, the NLSQ merit function in a metric based on $[\mathbf{J}(\mathbf{x}_k)^T\mathbf{J}(\mathbf{x}_k)]^{-1}$; see Bock et al. [2000]. For a comprehensive discussion of this traditional approach to nonlinear equation-solving, as summarized in Figure 1.1, see Dennis and Schnabel [1983].

In this chapter we consider two other methods for solving the nonlinear equation-solving problem that are closer to the modern portrait of Figure 1.2. The first, the Euler–Newton (EN) homotopy, or path-following, method, is well known and has been studied in depth in the literature. It is reviewed only very briefly here. The second, the Lagrange–Newton–Cauchy

[1]Here "global algorithm" means an algorithm that is often able to converge to a *local* minimum of F from a starting point that is far away from a solution of the associated system of equations, and it should not be confused with a GUM algorithm that is designed to find a *global minimum* of F; see Chapter 1 and, in particular, Figure 1.2.

(LNC) method based on Lagrangian-type potential functions, is relatively new and is considered in more detail.

3.2 The EN Method

An approach to solving an NEQ problem that makes sound common sense is to seek a "neighboring" system of equations whose solution is easy to obtain, or even directly at hand, and then to retrace a "guiding path" that leads from this known solution back to the solution of the original system. The underlying *homotopy principle*, which has historical roots in venerated works of Henri Poincaré, formalizes this approach. The idea is to introduce a parameter t into the system so that for some distinguished value, typically $t = 0$, the parameterized system has a known solution, and for some other distinguished value, typically $t = 1$, the parameterized system is identical to the original system of equations. By continuously changing the introduced parameter from 0 to 1, a path of solutions can be traced, leading from the known initial solution to a solution of the original problem.

Thus, consider the system $\mathbf{h}(\mathbf{x}) = \mathbf{0}$ as in (3.1) with solution $\mathbf{x}^*$. Given any point $\mathbf{x}^{(0)}$, there is an obvious and very natural way to introduce a parameter t so that the parameterized system fulfills the foregoing requirements, namely,

$$\mathbf{H}(\mathbf{x}, t) \equiv \mathbf{h}(\mathbf{x}) - (1 - t)\mathbf{h}(\mathbf{x}^{(0)}) = \mathbf{0}. \tag{3.9}$$

This parametrization and associated mapping $\mathbf{H}$ is termed the *global homotopy*, and it has been studied in detail from both theoretical and algorithmic standpoints. See, in particular, Allgower and Georg [1990], Garcia and Zangwill [1981], Keller [1978], [1987], Hirsch and Smale [1979], and references cited therein. Other useful references are Chow, Mallet-Paret and Yorke [1978], Ortega and Rheinboldt [1970], and Watson [1986]. The monograph of Garcia and Zangwill [1981] is an especially readable overview of the field.

Here, we give a *brief and informal review* of some known results whose purpose is twofold:

- to highlight the fact that the global homotopy and the EN method lie at the foundation of NEQ solution techniques;
- to lay the basis for subsequent usage in the LP setting in Chapter 10.

For the latter purpose, it is more convenient to define the homotopy system using the parameter $\mu \equiv (1 - t)$, which thus *decreases* to 0 as a path is traced to a solution. The global homotopy then becomes

$$\mathbf{H}(\mathbf{x}, \mu) \equiv \mathbf{h}(\mathbf{x}) - \mu\mathbf{h}(\mathbf{x}^{(0)}) = \mathbf{0}, \tag{3.10}$$

where $\mu \in [0, 1]$. *Henceforth, we will use the μ parameterization.*

In order to maintain continuity with the formulation of homotopy-based LP algorithms of Chapter 10, we will use the alternative *superscript notation for iterates of an algorithm* in the present section, for example, $\mathbf{x}^{(k)}$, $k = 0, 1, 2, \ldots$. (See also the comments on notation on page xvii.) Similarly, when the homotopy parameter μ is iterated, successive values are denoted by $\mu^{(k)}$, $k = 0, 1, 2, \ldots$, with $\mu^{(0)} = 1$.

3.2.1 Definitions and Path Existence

The mapping $\mathbf{h}$ that defines the original system is often restricted so as to inhabit a certain domain, say D, with boundary ∂D and interior D^0. The parameter μ can cover the real line, but commonly it is restricted to $U = \{\mu | 0 \leq \mu \leq 1\}$. Correspondingly, $\mathbf{H}$ is restricted to $D \times U$.

Define

$$\mathbf{H}^{-1}(\mu) = \{\mathbf{x} | \mathbf{H}(\mathbf{x}, \mu) = \mathbf{0},\ \mathbf{x} \in D\}.$$

Let $\mathbf{H}'(\mathbf{x}, \mu)$ denote the $n \times (n+1)$ Jacobian matrix of partial derivatives of the global homotopy mapping $\mathbf{H} : R^{n+1} \to R^n$. Denote its first n columns by $\mathbf{H}'_x(\mathbf{x}, \mu)$, the $n \times n$ Jacobian matrix corresponding to the original mapping $\mathbf{h}$, and its last column by $\mathbf{H}'_\mu(\mathbf{x}, \mu)$, the n-vector corresponding to the partial derivative of $\mathbf{H}(\mathbf{x}, \mu)$ with respect to μ.

The global homotopy is said to be *regular at the point* μ if $\mathbf{H}'_x(\mathbf{x}, \mu)$ is of full rank for all $\mathbf{x} \in \mathbf{H}^{-1}(\mu)$.

Furthermore, the global homotopy is said to be *boundary-free at the point* μ if $\mathbf{x} \in D^0$ for all $\mathbf{x} \in \mathbf{H}^{-1}(\mu)$.

Suppose $\mathbf{h}$ is twice continuously differentiable, the set D is compact, and the global homotopy $\mathbf{H}$ is regular and boundary-free for all points $\mu \in \overline{U}$, where $\overline{U} = \{\mu | 0 < \mu \leq 1\}$. These conditions are *sufficient* to ensure the existence, via the implicit function theorem, of a *unique differentiable path of solutions* $(\mathbf{x}(\mu), \mu)$ that leads from the initial point $(\mathbf{x}^{(0)}, 1)$ to a solution of the original system $(\mathbf{x}^*, 0)$. Note also that the path lies in the interior of D, but can terminate on its boundary. The above regularity assumption implies that the path cannot turn back on itself as μ decreases monotonically.

The foregoing homotopy path lies in a space of dimension $n + 1$, and its projection into the space of the $\mathbf{x}$-variables, which is of dimension n, will be called the *lifted* path. In the discussion that follows we will usually use the term "path" and allow the context to determine whether reference is being made to the original homotopy path in R^{n+1} or the lifted path in R^n.

The matrix $\mathbf{H}$ is said to be *regular* if the $n \times (n + 1)$ Jacobian matrix $\mathbf{H}'(\mathbf{x}, \mu)$ is of full rank for all $(\mathbf{x}, \mu) \in \mathbf{H}^{-1}$, where $\mathbf{H}^{-1} = \{(\mathbf{x}, \mu) | \mathbf{H}(\mathbf{x}, \mu) = \mathbf{0}, \mathbf{x} \in D, \mu \in U\}$. Under this weaker assumption, the homotopy path remains well-defined, but it can have turning points. (In fact, the regularity assumption can be totally eliminated as discussed in Garcia and Zangwill

[1981] or in one of the other monographs cited above.) The "regularity at a point" assumption, however, *suffices for the discussion of path-following* in the next three subsections and for purposes of application in the LP setting in Chapter 10.

Also, for later reference, we note the following:

- The homotopy path can be characterized by a *homotopy differential equation (HDE)* system. This is obtained by differentiating the homotopy equations $\mathbf{H}(\mathbf{x}(\mu), \mu) = \mathbf{0}$ with respect to μ, and is given as follows:
$$\frac{d\mathbf{x}}{d\mu} = -[\mathbf{H}'_x(\mathbf{x}, \mu)]^{-1}[\mathbf{H}'_\mu(\mathbf{x}, \mu)]. \tag{3.11}$$
The initial condition of this system of ordinary differential equations is $\mathbf{x}(1) = \mathbf{x}^{(0)}$.
- Consider any point $\mathbf{x}(\mu)$ on the path defined by the global homotopy (3.10) and any vector $\mathbf{z} \in R^n$ such that $\mathbf{z}^T\mathbf{h}(\mathbf{x}^{(0)}) \neq 0$. Then
$$\mu = \frac{\mathbf{z}^T\mathbf{h}(\mathbf{x}(\mu))}{\mathbf{z}^T\mathbf{h}(\mathbf{x}^{(0)})}. \tag{3.12}$$
Since $\mathbf{h}(\mathbf{x}^{(0)}) \neq \mathbf{0}$, a convenient choice is $\mathbf{z} = \mathbf{h}(\mathbf{x}^{(0)})$.
- In the discussion of path-following algorithms in the next three subsections, we will denote points that lie on the (lifted) path by $\mathbf{x}(\mu^{(k)})$, $k = 0, 1, 2, \ldots$. Under the foregoing regularity assumptions, each such point has a unique value of the homotopy parameter that defines it, namely, $\mu = \mu^{(k)}$.

  Algorithm iterates that lie *off* the path will be denoted by $\bar{\mathbf{x}}^{(k)}$, $k = 0, 1, 2, \ldots$. A point $\bar{\mathbf{x}}^{(k)}$ will often have a particular value of the homotopy parameter *associated* with it. This association between iterate and parameter value will usually be stated verbally in the discussion below. But if a mathematical notation becomes necessary, we will use $\bar{\mathbf{x}}[\mu^{(k)}]$. The square brackets highlight the fact that the point is not defined as a *function* of μ, as is the case with points that lie on the path.

3.2.2 Embedding Algorithms

Given any point on the path, say $\mathbf{x}(\mu^{(k)})$, reduce the value of the parameter from $\mu^{(k)} > 0$ to $\mu^{(k+1)} > 0$, and apply Newton's method to the nonlinear system $\mathbf{H}(\mathbf{x}, \mu^{(k+1)}) = \mathbf{0}$ from the starting point $\mathbf{x}(\mu^{(k)})$. The *conceptual*[2]

---

[2]The distinction between *conceptual* and *implementable* algorithms is due to Polak [1971]. Our brief overview, in this subsection and the next, focuses on some key characteristics of conceptual and implementable embedding and EN algorithms, and it does *not* address important details concerning the permissible change in homotopy parameter and choice of predictor and corrector step sizes that will guarantee convergence.

version of the method solves this system *exactly* to obtain the point on the path $\mathbf{x}(\mu^{(k+1)})$ corresponding to the new parameter value $\mu^{(k+1)}$.

The *implementable* version uses only a *small* number of iterations T of Newton's method (e.g., $T = 1$), which is initiated from an *approximation* to $\mathbf{x}(\mu^{(k)})$, say $\overline{\mathbf{x}}^{(k)}$, with associated parameter value $\mu^{(k+1)}$ and in turn yields an approximation to $\mathbf{x}(\mu^{(k+1)})$, say $\overline{\mathbf{x}}^{(k+1)}$. The procedure is repeated at this new point $\overline{\mathbf{x}}^{(k+1)}$. If $\overline{\mathbf{x}}^{(k)}$ is a "sufficiently accurate" approximation to $\mathbf{x}(\mu^{(k)})$ and the parameter value $\mu^{(k)}$ is not changed "too much," one would intuitively expect the starting point $\overline{\mathbf{x}}^{(k)}$ to remain within the domain of quadratic convergence when Newton's method is applied to the system $\mathbf{H}(\mathbf{x}, \mu^{(k+1)}) = \mathbf{0}$. A single iteration of this so-called Newton corrector can then be expected to yield an approximation $\overline{\mathbf{x}}^{(k+1)}$ that is a sufficiently accurate approximation to $\mathbf{x}(\mu^{(k+1)})$, the true solution of $\mathbf{H}(\mathbf{x}, \mu^{(k+1)}) = \mathbf{0}$. This, in turn, would ensure that the procedure can be repeated.

The foregoing *embedding* algorithms[3] are especially useful for purposes of theoretical analysis, as will be seen later in Chapter 10.

3.2.3 EN Algorithms

Embedding has given way to modern homotopy techniques that are much more effective in practice. They are premised on the HDE system (3.11) and draw on the spectrum of techniques available for solving this system of ordinary differential equations, *but always within the context of the parameterized system that gave rise to the HDE.*

The most basic ordinary differential equations (ODE) technique of this type is Euler's method. Its associated Euler direction at any point is obtained by locally linearizing the HDE system (3.11). At a point that lies *on* the path, this direction is along the tangent to the path. At a point *off* the path, the Euler direction derived from the HDE (3.11) is along the tangent to a *different* path, defined by the same HDE, but with a *modified initial condition.* Paths of the latter type form a "tube" around the original homotopy path defined by the HDE and the initial condition $\mathbf{x}(1) = \mathbf{x}^{(0)}$.

Effective procedures can be obtained by treating the Euler direction as a predictor and augmenting it with a Newton corrector. Again one can distinguish between conceptual and implementable versions. Let us consider the former.

Given a point on the path $\mathbf{x}(\mu^{(k)})$, take a nonzero step along the Euler predictor. Denote the new point by $\overline{\mathbf{x}}^{(k+1)}$ and the associated parameter value by $\mu^{(k+1)}$, where the change $(\mu^{(k)} - \mu^{(k+1)})$ in the homotopy parameter, which can be used to define the length of the predictor step, is constrained by $\overline{\mathbf{x}}^{(k+1)} \in D$. Apply Newton's method—the Newton corrector—

---

[3]They are sometimes called vertical-predictor/horizontal Newton-corrector algorithms; see Garcia and Zangwill [1981].

to the system $\mathbf{H}(\mathbf{x}, \mu^{(k+1)}) = \mathbf{0}$ with starting point $\overline{\mathbf{x}}^{(k+1)}$ to obtain the solution $\mathbf{x}(\mu^{(k+1)})$. Then repeat the procedure for $k = 0, 1, 2, \ldots$.

In the *implementable* version, form the Euler predictor direction at a point $\overline{\mathbf{x}}^{(k)}$ with associated parameter value $\mu^{(k)}$; i.e., a point $\overline{\mathbf{x}}[\mu^{(k)}]$, in the notation introduced at the end of Section 3.2.1, that approximates $\mathbf{x}(\mu^{(k)})$. Proceed as in the conceptual version, but after taking the step along the Euler predictor to $\overline{\mathbf{x}}^{(k+1)}$, use only a *finite* number of Newton corrector iterations—ideally a single iteration—to obtain an updated solution, which approximates $\mathbf{x}(\mu^{(k+1)})$ and is denoted again by $\overline{\mathbf{x}}^{(k+1)}$. Then repeat the procedure for $k = 0, 1, 2, \ldots$.

3.2.4 Algorithmic Enhancements

The homotopy differential equation (HDE) is a means to an end, and it is important that the homotopy approach *not be equated* to this particular *initial value ordinary differential equation (ODE) problem* and solved directly using general-purpose ODE techniques or off-the-shelf ODE software. To do so is to "cripple the homotopy method from the outset," in an apt phrase of Allgower [1998].

The following two quotations[4] highlight the underlying, fundamental issue, albeit in somewhat repetitive language. The first is from Garcia and Zangwill [1981, p. 300] as follows:

> A *terminal value problem*, which is the type we are often interested in, is usually easier to solve than an initial value problem.... The reason is that the intial value problem ... uses only differential equation information. However, the terminal value problem employs the homotopy **H** and the differential equation .... In essence, the homotopy permits us to correct and restart.... The initial value problem ... uses only the differential equations and no homotopy information, so has no way to correct itself. The homotopy, which is used for the terminal value but not the initial value problem, provides a great deal of additional data and permits the correction.

The second quotation is from Watson [1986]:

> One noteworthy difference between general curve tracking algorithms and homotopy curve tracking algorithms is the objective: the curve itself for general curve tracking and the solution at the end of the curve for homotopy algorithms. This means that the curve itself is not important and sophisticated homotopy algorithms (as in HOMPACK, e.g.) actually *change* the curve that

[4]Italics in these quotations are ours.

> is being tracked as they proceed. This strategy has a rigorous theoretical justification, since changing the curve amounts to changing the parameter vector in the homotopy map, and for almost all parameter vectors the zero curve of the homotopy map is guaranteed to reach a solution. Furthermore, homotopy algorithms are inherently stable because all zero curves are confluent at the solution—the ideal situation where "all roads lead to Rome." However, following the zero curve γ too loosely can be disastrous ... so there is a delicate balance between efficiency and robustness. This balance needs further study, *perhaps leading to fundamentally different algorithms.*

These quotations highlight the fact that there is no fundamental requirement in the homotopy method to stay within a *tight tubular neighborhood* of the original homotopy path and to follow the latter accurately. A wide variety of EN algorithmic enhancements are possible, which are now outlined.

Restart Strategies

The homotopy can be restarted, for example, after a corrector phase, which is tantamount to *changing* the underlying homotopy differential equation. Consider any iterate, say $\overline{\mathbf{x}}^{(k)}$, obtained from one of the foregoing implementable path-following strategies. In principle, this iterate can be any point in the domain D. Two common ways of restarting the homotopy at $\overline{\mathbf{x}}^{(k)}$ are as follows (Garcia and Zangwill [1981 p. 297–300]):

1. Restart homotopy over the range $\mu \in [0, \mu^{(k)}]$ with given $\mu^{(k)} > 0$:

$$\overline{\mathbf{H}}(\mathbf{x}, \mu) \equiv \mathbf{H}(\mathbf{x}, \mu) - \frac{\mu}{\mu^{(k)}}\mathbf{H}(\overline{\mathbf{x}}^{(k)}, \mu^{(k)}) = \mathbf{h}(\mathbf{x}) - \frac{\mu}{\mu^{(k)}}\mathbf{h}(\overline{\mathbf{x}}^{(k)}), \quad (3.13)$$

2. Restart homotopy over the range $\mu \in [0, 1]$ with $\mu^{(k)}$ reset to 1:

$$\overline{\mathbf{H}}(\mathbf{x}, \mu) \equiv \mathbf{H}(\mathbf{x}, \mu) - \mu\mathbf{H}(\overline{\mathbf{x}}^{(k)}, 1) = \mathbf{h}(\mathbf{x}) - \mu\mathbf{h}(\overline{\mathbf{x}}^{(k)}), \quad (3.14)$$

where the last expression in each restart strategy above follows directly from expression (3.10). Observe that the only difference between the two restarting strategies is the range of the homotopy parameter. The first retains the value of the homotopy parameter where the previous homotopy was discarded, and the second corresponds to a restart of the homotopy at $\overline{\mathbf{x}}^{(k)}$ with initial $\mu = 1$.

When the restart iterate lies *on* the original homotopy path; i.e., when the restart is performed at a point $\overline{\mathbf{x}}[\mu^{(k)}] \equiv \mathbf{x}(\mu^{(k)})$, for some $\mu^{(k)} > 0$, then the restarted homotopy path must be *identical* to the original homotopy path. The first restart strategy then leaves the homotopy unchanged, and the second strategy corresponds to a *reparameterization* of the homotopy.

Predictor–Corrector Variants

Many variants are possible. For example, several predictor steps can be taken prior to a corrector step, or the corrector phase can be entirely eliminated; i.e., only the Euler predictor step can be taken (with a suitable choice of initial step length coupled with a backtracking strategy) and then followed by a restart at the point $\overline{\mathbf{x}}^{(k+1)}$. This latter strategy is the so-called *global Newton* method; see, Smale [1976], Goldstein [1991].

The Newton iteration, in the form described in Sections 3.2.2 and 3.2.3, is called a *horizontal* Newton corrector (Garcia and Zangwill [1981]), because the homotopy parameter is held constant at value $\mu = \mu^{(k+1)}$. Consider the space of variables $(\mathbf{x}, \mu)$ and the hyperplane in R^{n+1} given by

$$\eta^T \left[(\mathbf{x}, \mu) - \left(\overline{\mathbf{x}}^{(k+1)}, \mu^{(k+1)}\right)\right] = 0, \tag{3.15}$$

where $\eta \in R^{n+1}$ is the normal vector to the hyperplane. This vector is defined by $\eta^T \equiv \mathbf{e}_{n+1}^T = (\mathbf{0}^T, 1)$, where $\mathbf{0} \in R^n$ is a vector of zeros. The horizontal Newton corrector can then be viewed as applying Newton's method to the following system of $(n+1)$ equations in an $(n+1)$-dimensional space of variables $(\mathbf{x}, \mu)$:

$$\begin{aligned} \mathbf{H}(\mathbf{x}, \mu) &= \mathbf{0} \\ \eta^T [(\mathbf{x}, \mu) - \left(\overline{\mathbf{x}}^{(k+1)}, \mu^{(k+1)}\right)] &= 0. \end{aligned} \tag{3.16}$$

The procedure is initiated from the point $\left(\overline{\mathbf{x}}^{(k+1)}, \mu^{(k+1)}\right)$, and all Newton iterates must obviously lie in the hyperplane.

The alternative formulation permits an extension of the approach by employing a different normal vector η; again see Garcia and Zangwill [1981]. This can be chosen "artfully" so that the associated hyperplane intersects the path at an improved point; i.e., one that lies further along the path. The Newton corrector is applied as above to the system (3.16) with the alternative choice of the vector η. Again, all Newton iterates must lie in the hyperplane. Another variant is obtained by retaining the choice $\eta^T \equiv \mathbf{e}_{n+1}^T$ and *shifting* the horizontal hyperplane by making a change in the right-hand side in the second equation of (3.16) to $\left(\sigma^{(k)} - 1\right)\mu^{(k+1)}$ with $\sigma^{(k)} < 1$. This equation is equivalent to the equation $\mu = \sigma^{(k)}\mu^{(k+1)}$. We will see an example of shifted hyperplanes within interior-point LP algorithms in Chapter 10. Further variants are obtained by changing both the normal vector η and the right-hand side in (3.16).

Bipath-Following Strategies

A fundamental EN enhancement[5] is to use a *restarted* homotopy path to define the Euler predictor and the *original* path to guide the Newton

[5]To date, this enhancement has not been explored in any detail in the general nonlinear equation-solving setting.

corrector. In other words, *two* HDE systems are used simultaneously, one to define the predictor and a second (and different) HDE to define the corrector. A bipath-following algorithm can obviously also be *initiated* at a point $\bar{\mathbf{x}}[1]$ that lies *off* the path used to guide the Newton corrector.

We shall see that the foregoing enhancements in combination—restarts, shifted Newton correctors, bipath-following—play a key role in the formulation of interior-point LP algorithms in Chapter 10. *Issues of convergence to a desired solution must be addressed when a single path is followed, conceptually or implementably, and when bipath and restart strategies are used,* as will be discussed in the linear programming setting. We will return also to the broader philosophy of homotopy techniques in the last section of Chapter 10.

3.3 The LNC Method

In order to shed further light on difficulties with using the sum-of-squares merit function to measure progress in the implementation of Newton's method, let us now introduce the idea of a potential function associated with the mapping $\mathbf{h}$ in (3.1) and give conditions for its existence.

Definition: A mapping $\mathbf{h} : R^n \to R^n$ is a gradient (or potential) mapping if there exists a differentiable *potential function* $p : R^n \to R$ such that $\mathbf{h}(\mathbf{x}) = \nabla p(\mathbf{x})$ for all $\mathbf{x} \in R^n$.

Symmetry Principle: Let $\mathbf{h} : R^n \to R^n$ be continuously differentiable. Then $\mathbf{h}$ is a gradient mapping if and only if the Jacobian matrix of $\mathbf{h}$ is symmetric for all $\mathbf{x} \in R^n$. If it exists, the potential function p as defined above can be explicitly constructed by

$$p(\mathbf{x}) = \int_0^1 (\mathbf{x} - \bar{\mathbf{x}})^T \mathbf{h}(\bar{\mathbf{x}} + t(\mathbf{x} - \bar{\mathbf{x}}))dt, \tag{3.17}$$

where $\bar{\mathbf{x}} \in R^n$ is an arbitrarily selected point.

Proof: See Ortega and Rheinboldt [1970, Section 4.1, pp. 95–96]. □

Note also that the foregoing construction of p is closely related to a standard mean value theorem and also to the fundamental theorem of calculus. Strang [1986] gives a good discussion of the importance of a potential function and, in particular, makes the following observation (Section 3.4, page 199):

> When there is a potential function, the problems of engineering and science are enormously simplified .... It is the same in physics and biology and chemistry; if there is a potential func-

> tion then the flow is governed by its derivative or its gradient or its nodal differences. Equilibrium comes from the drive to equalize potentials....

We now see the root of the difficulty with the NLSQ approach to solving the NEQ problem (3.1). In general, the Jacobian matrix of the mapping $\mathbf{h}$ in (3.1) is *nonsymmetric*, so that it cannot have an associated potential function. Thus, when the sum-of-squares function F in (3.3) is used as a merit function and its critical points are sought, there is no guarantee that such points are solutions of the system of equations; i.e., there is an inherent flaw in using F, as we have seen in the examples at the beginning of this chapter.

With the foregoing in mind, we will seek a system of nonlinear equations *that subsumes the system (3.1) and has an associated potential function.* This approach, in its broad concept, but obviously not in its specific details, is motivated by the interior LP technique pioneered by Karmarkar [1984], where an appropriate potential function is associated with a linear program to facilitate its solution (see Chapter 12). The use of potential functions will be a *recurrent theme* throughout this monograph.

3.3.1 An Alternative View

We adopt an alternative view of the nonlinear equation-solving problem that is consistent with the portrait of differentiable programming given in Figure 1.2. Instead of treating nonlinear equation-solving as a subject closely allied to *unconstrained optimization*—the traditional approach summarized in Figure 1.1 and discussed, for example, in Dennis and Schnabel [1983]—let us view equation-solving as belonging to "the other half" of differentiable programming; i.e., let us treat the equations in (3.1) as the *equality constraints* of a nonlinear optimization problem whose objective function is unspecified and therefore open to choice. In this setting, "the first half"; see, again Figure 1.2—contains and is premised on unconstrained optimization, where an objective function is given and either no constraints are specified, or an ellipsoidal trust region (Section 2.2.1) or even a set of bound constraints is introduced, artificially and purely as a convenience, in order to facilitate the search for the optimal point. *This alternative view of nonlinear equation-solving is very natural*, and it is indeed somewhat surprising that it has not been more fully exploited to date as an avenue for developing effective algorithms.

To be more precise, consider the general equality-constrained optimization problem NECP of Chapter 1, expression (1.6). We will view the system (3.1) as a particular problem of this type with $m = n$ and an objective function f that can be chosen to facilitate the search for a solution. For example, define the artificial objective function to be $\frac{1}{2}(\mathbf{x} - \hat{\mathbf{x}})^T(\mathbf{x} - \hat{\mathbf{x}})$, where $\hat{\mathbf{x}}$ is a given point, typically the zero vector or a recent iterate.

When the solution of (3.1) is not unique, this imposes a requirement that the selected solution should be as close as possible to $\hat{\mathbf{x}}$. The reformulated problem is as follows:

$$\begin{aligned} &\text{minimize } f(\mathbf{x}) = \tfrac{1}{2}(\mathbf{x} - \hat{\mathbf{x}})^T(\mathbf{x} - \hat{\mathbf{x}}) \\ &\text{s.t. } \mathbf{h}(\mathbf{x}) = \mathbf{0}. \end{aligned} \tag{3.18}$$

The Lagrange or Karush–Kuhn–Tucker (KKT) necessary conditions for optimality of (3.18) are as follows:

$$\begin{aligned} \nabla_x L(\mathbf{x}, \lambda) &= \nabla f(\mathbf{x}) + \mathbf{J}(\mathbf{x})^T \lambda = \mathbf{0}, \\ \nabla_\lambda L(\mathbf{x}, \lambda) &= \mathbf{h}(\mathbf{x}) = \mathbf{0}, \end{aligned} \tag{3.19}$$

where L is the Lagrangian function of the foregoing problem (3.18), namely,

$$L(\mathbf{x}, \lambda) = f(\mathbf{x}) + \lambda^T \mathbf{h}(\mathbf{x}) \tag{3.20}$$

and $\lambda \in R^n$ is the associated vector of Lagrange multipliers. The larger system of nonlinear equations (3.19) has a symmetric Jacobian, and the Lagrangian function is its associated potential function. Put another way, the original system $\mathbf{h}(\mathbf{x}) = \mathbf{0}$ has been embedded in a higher-dimensional system (3.19) that has features capable of facilitating its solution, in particular, potential functions derived from the Lagrangian function (3.20) and its augmentations.

In optimization, attaining feasibility is often considered an "easy" task that involves finding a point in a large, continuous region of R^n. In problem (3.18), on the other hand, we are dealing with an opposite situation, where the feasibility task itself is "hard." Thus, it is important to note that we are *not* advocating here the naive approach of handing over the foregoing reformulation (3.18) to a standard, general-purpose constrained optimization library routine. Indeed, if this is attempted, using specifically the example (3.5) and the IMSL routine DNCONF, one finds that the routine does occasionally return with the solution near $(1.3, 1)$. But more often than not, it terminates with a message like, "Line search took more than 5 function calls," indicating that an unsatisfactory search direction has been generated. This failure to converge can be alleviated by increasing the default from 5, but the difficulty will not be eliminated; see also the comments at the end of Section 3.3.3.

We return, instead, to the *fundamentals of Lagrangian-based methods* for the foregoing problem (3.18)—a nonlinear equality-constrained problem with an artificial convex quadratic objective—in order to derive potential functions that are better suited to nonlinear equation-solving than the sum-of-squares merit function. In approaching this task, we can draw on a rich and very well developed theory and practice of Lagrangian-based techniques for constrained nonlinear optimization; see, in particular, Hestenes [1969],

Powell [1969], Rockafellar [1976], and others cited in the comprehensive overview of Bertsekas [1982].

3.3.2 Lagrangian Globalization

A critical point of the Lagrangian function associated with (3.19) is a saddle point. Although it yields a solution of the original system of equations, the function L is not well suited for direct use in an NC *minimization* algorithm, and there are well-known advantages to working instead with augmentations of the Lagrangian. We will choose one of the *most general augmentations*, the so-called differentiable exact penalty function of Di Pillo and Grippo [1979] (see also Bertsekas [1982], [1999]), namely,

$$P(\mathbf{x}, \lambda; c, \alpha) = L(\mathbf{x}, \lambda) + \tfrac{1}{2}c\|\mathbf{h}(\mathbf{x})\|^2 + \tfrac{1}{2}\alpha\|\nabla_x L(\mathbf{x}, \lambda)\|^2, \tag{3.21}$$

where $c \in R_+$ and $\alpha \in R_+$, $\|.\|$ denotes the Euclidean norm, and $\nabla_x L$ is the gradient vector with respect to the $\mathbf{x}$-variables of the Lagrangian function L. (Later, in Section 3.3.4, simpler and more practical variants of (3.21) are considered.)

The components of the gradient vector $\nabla P(\mathbf{x}, \lambda; c, \alpha)$ of P are easily derived and are given by

$$\nabla_x P(\mathbf{x}, \lambda; c, \alpha) = \nabla_x L(\mathbf{x}, \lambda) + c\mathbf{J}(\mathbf{x})^T\mathbf{h}(\mathbf{x}) + \alpha\nabla^2_{xx}L(\mathbf{x}, \lambda)\nabla_x L(\mathbf{x}, \lambda) \tag{3.22}$$

and

$$\nabla_\lambda P(\mathbf{x}, \lambda; c, \alpha) = \mathbf{h}(\mathbf{x}) + \alpha\mathbf{J}(\mathbf{x})\nabla_x L(\mathbf{x}, \lambda), \tag{3.23}$$

where $\mathbf{J}(\mathbf{x})$ denotes the Jacobian matrix of $\mathbf{h}$ at $\mathbf{x}$.

Analogously to the KKT system (3.19), which corresponds to critical points of L, the augmented system now under consideration is given by the critical points of P, namely,

$$\nabla_x L(\mathbf{x}, \lambda) + c\mathbf{J}(\mathbf{x})^T\mathbf{h}(\mathbf{x}) + \alpha\nabla^2_{xx}L(\mathbf{x}, \lambda)\nabla_x L(\mathbf{x}, \lambda) = \mathbf{0}, \tag{3.24}$$

$$\mathbf{h}(\mathbf{x}) + \alpha\mathbf{J}(\mathbf{x})\nabla_x L(\mathbf{x}, \lambda) = \mathbf{0}, \tag{3.25}$$

or in other words, (3.21) is the potential function of the foregoing augmented system.

The correlation between solutions of the augmented system; i.e., critical points of P, and solutions of the original system (3.1) is, of course, a key consideration that must be addressed. In the following discussion, let us assume that the Jacobian matrix $\mathbf{J}(\mathbf{x}^*)$ is nonsingular at any *solution* point $\mathbf{x}^*$ of (3.1). (For purposes of present discussion, we can even make the simplifying assumption that the solution point $\mathbf{x}^*$ is unique.) Elsewhere, of course, the Jacobian matrix may be singular. The next set of propositions demonstrates, in essence, that a solution point of the original system (3.1) has an associated critical point of P. And within a compact set that can be defined by bound constraints and made as large as desired, a critical point of P is a Karush–Kuhn–Tucker (KKT) point of (3.18), and its $\mathbf{x}$-component thus yields a solution of the original system, provided c is chosen to be sufficiently large and a correlated choice is made for α that is sufficiently small; i.e., spurious critical points of P are "pushed out to infinity" for appropriate choices of c and α. A desired critical point, corresponding to a solution of the NEQ problem, can be reached by a *descent* procedure. And an isolated solution of the original system corresponds to an isolated local *minimum* of P when additionally $\alpha > 0$; i.e., a standard minimizing algorithm can be employed.

These propositions build on the development of Bertsekas [1982], and proofs can be found in Nazareth and Qi [1996]. For the interested reader, we mention that there is considerable room for strengthening these results. These propositions, whose content was summarized in the previous paragraph, can be *skipped at a first reading* without any loss of continuity.

Proposition 3.1: Suppose $\overline{\mathbf{x}}$ is a critical point of $F(\mathbf{x}) = \frac{1}{2}\mathbf{h}(\mathbf{x})^T\mathbf{h}(\mathbf{x})$ but is not a solution of (3.1). Then, for any choice of λ, the point $(\overline{\mathbf{x}}, \lambda)$ is *not* a critical point of P.

Proposition 3.2: Suppose $\mathbf{x}^* \in R^n$ is a solution of (3.1) and $\mathbf{J}(\mathbf{x}^*)$ is nonsingular. Let $\lambda^* = -\mathbf{J}(\mathbf{x}^*)^{-T}\nabla f(\mathbf{x}^*)$. Then $(\mathbf{x}^*, \lambda^*)$ is a critical point of P for any $c \in R_+$ and $\alpha \in R_+$.

Proposition 3.3: Consider any point $\overline{\mathbf{x}}$ with $\mathbf{h}(\overline{\mathbf{x}}) \neq \mathbf{0}$ and the critical point $(\mathbf{x}^*, \lambda^*)$ as in Proposition 3.2. Given compact sets X and Λ for which $\overline{\mathbf{x}} \in X$, $\mathbf{x}^* \in X$, and $\lambda^* \in \Lambda$, let $|f(\mathbf{x})| \leq U_f$, $\mathbf{x} \in X$, and let c_0 be a positive constant such that for any $\mathbf{x} \in X$ and $\lambda \in \Lambda$,

$$c_0 \geq \frac{2\left(|L(\mathbf{x}, \lambda)| + \frac{1}{10}\|\nabla_x L(\mathbf{x}, \lambda)\|^2 + U_f\right)}{\|\mathbf{h}(\overline{\mathbf{x}})\|^2}. \tag{3.26}$$

Then

$$P(\mathbf{x}^*, \lambda^*; c, \alpha) < P(\overline{\mathbf{x}}, \lambda; c, \alpha)$$

for any $\lambda \in \Lambda$, $\alpha \in R_+$, and $c > c_0$.

Proposition 3.4: Suppose that $(\tilde{\mathbf{x}}, \tilde{\lambda}) \in X \times \Lambda$ is any critical point of P. Suppose $c > c_0$, with c_0 defined by (3.26). Let α be chosen to satisfy

$$0 \leq c\alpha \leq b, \tag{3.27}$$

where b is a nonnegative constant such that, for any $\mathbf{x} \in X$ and $\lambda \in \Lambda$,

$$b \leq \tfrac{1}{5}c_0; \quad b\|\nabla^2_{xx} L(\mathbf{x}, \lambda)\| \leq \tfrac{1}{3}c_0; \quad \text{and} \quad b\|\mathbf{J}(\mathbf{x})^T\mathbf{J}(\mathbf{x})\| \leq \tfrac{1}{3}. \tag{3.28}$$

Then $\tilde{\mathbf{x}}$ satisfies $\mathbf{h}(\tilde{\mathbf{x}}) = \mathbf{0}$.

Proposition 3.5: Let $\mathbf{x}^*$ be an isolated solution of (3.1) with $\mathbf{J}(\mathbf{x}^*)$ nonsingular. Define $X^* \subset R^n$ to be a compact set such that $\mathbf{x}^*$ lies in its interior and $\mathbf{J}(\mathbf{x})$ is nonsingular for all $\mathbf{x} \in X^*$. (Such a set clearly exists by continuity of $\mathbf{J}(\mathbf{x})$ at $\mathbf{x}^*$.) Let $\Lambda^* \subset R^n$ be a compact set such that λ^* defined as in Proposition 3.2 lies in the interior of Λ^*. Let c and $\alpha > 0$ be chosen to satisfy the requirements of Propositions 3.3 and 3.4 for the current choice $X = X^*$ and $\Lambda = \Lambda^*$. Then $(\mathbf{x}^*, \lambda^*)$ is the unique minimum of P over $X^* \times \Lambda^*$.

3.3.3 An Illustrative Algorithm

We will now formulate an illustrative LNC algorithm based on the foregoing ideas by using a simple Newton–Cauchy procedure to minimize P in the $(\mathbf{x}, \lambda)$-space. Consider a steepest-descent search direction $\mathbf{d}_k$ at an iterate, say $(\mathbf{x}_k, \lambda_k)$, which is defined by the negative gradient vector (3.22)–(3.23) as follows:

$$\mathbf{d}_k = -\begin{bmatrix} \nabla_x P(\mathbf{x}_k, \lambda_k; c, \alpha) \\ \nabla_\lambda P(\mathbf{x}_k, \lambda_k; c, \alpha) \end{bmatrix}. \tag{3.29}$$

Note that the last term in (3.22) does not require second derivatives of the Lagrangian function, because it can be obtained by a finite-difference operation along the direction $\nabla_x L(\mathbf{x}_k, \lambda_k)$ as follows:

$$\frac{\nabla_x L([\mathbf{x}_k + t\nabla_x L(\mathbf{x}_k, \lambda_k)], \lambda_k) - \nabla_x L(\mathbf{x}_k, \lambda_k)}{t},$$

where t is a small positive number, for example, the square root of the machine-precision constant (often denoted by *macheps*). Of course, this requires an extra Jacobian evaluation and makes the use of the potential P more expensive than the sum-of-squares merit function F.

Given $\sigma \in \left(0, \frac{1}{2}\right)$ and $\beta \in (0, 1)$, let the next iterate be defined by the well-known Goldstein–Armijo rule (see, for example, Kelley [1999] or Bertsekas [1999]), namely,

$$\begin{bmatrix} \mathbf{x}_{k+1} \\ \lambda_{k+1} \end{bmatrix} = \begin{bmatrix} \mathbf{x}_k \\ \lambda_k \end{bmatrix} + \beta^m \mathbf{d}_k, \tag{3.30}$$

where m is chosen in succession from the sequence $\{0, 1, 2, 3, \ldots\}$ until the following inequality is satisfied:

$$P(\mathbf{x}_k, \lambda_k; c, \alpha) - P(\mathbf{x}_{k+1}, \lambda_{k+1}; c, \alpha) \geq -\sigma\beta^m \mathbf{d}_k^T \nabla P(\mathbf{x}_k, \lambda_k; c, \alpha). \tag{3.31}$$

If we assume that all iterates remain within the compact set $X \times \Lambda$ of Proposition 3.3 of the previous subsection, then the standard theory of convergence of steepest descent, coupled with the foregoing properties of P, implies that $\mathbf{x}_k$ converges to a solution of (3.1).

In order to enhance the rate of convergence in a computational experiment described below, we will extend the foregoing algorithm by taking the search vector to be in the direction of steepest descent in the metric defined by a positive definite approximation to the Hessian matrix of P. Since considerations of numerical stability are not at issue here, we shall work with the inverse of this matrix, purely for reasons of convenience. This inverse is denoted by $\mathbf{W}_k$ at the current iterate k. Note that it is a $2n \times 2n$ matrix. This is initially taken to be the identity matrix, and thereafter it is updated using the symmetric rank-one update (see Section 2.2.2) as follows:

$$\mathbf{W}_{k+1} = \mathbf{W}_k + \frac{(\mathbf{s}_k - \mathbf{W}_k\mathbf{y}_k)(\mathbf{s}_k - \mathbf{W}_k\mathbf{y}_k)^T}{(\mathbf{s}_k - \mathbf{W}_k\mathbf{y}_k)^T\mathbf{y}_k}, \tag{3.32}$$

where

$$\mathbf{s}_k = \begin{bmatrix} \mathbf{x}_{k+1} - \mathbf{x}_k \\ \lambda_{k+1} - \lambda_k \end{bmatrix}, \quad \mathbf{y}_k = \nabla P(\mathbf{x}_{k+1}, \lambda_{k+1}; c, \alpha) - \nabla P(\mathbf{x}_k, \lambda_k; c, \alpha).$$

The update is performed only when the denominator in (3.32) is safely positive, which ensures that the updated matrix is positive definite and, in turn, yields a metric. The associated search direction vector $\mathbf{d}_k$ is now defined by

$$\mathbf{d}_k = -\mathbf{W}_k \begin{bmatrix} \nabla_x P(\mathbf{x}_k, \lambda_k; c, \alpha) \\ \nabla_\lambda P(\mathbf{x}_k, \lambda_k; c, \alpha) \end{bmatrix}. \tag{3.33}$$

The algorithm was implemented in Fortran-77 and run on the example (3.5)–(3.6). We took $f(\mathbf{x}) = \frac{1}{2}\mathbf{x}^T\mathbf{x}$ and used the following settings for the constants: $c = 10^3$, $\alpha = 10^{-5}$, $\sigma = 10^{-3}$, and $\beta = 0.1$. The table illustrates the results that were obtained for several starting points:

Starting Point	LGLOB	DNEQNJ
(100, 10, 0, 0)	S	F
$(-100, 10, 0, 0)$	S	F
(50, 10, 0, 0)	S	F
$(-50, 10, 0, 0)$	S	F
$(-10, 10, 0, 0)$	S	S
(500, 100, 0, 0)	S*	F*
$(-100, 1, 0, 0)$	S*	F*

In the table, LGLOB denotes the foregoing algorithm, and DNEQNJ is the IMSL equation-solving routine mentioned in the examples in the introduction to this chapter. For DNEQNJ, only the first two components of each starting point are used. Here S denotes a successful run that terminated at a good approximation to the solution, and F denotes a failure. In this particular experiment, we found that DNEQNJ failed most of the time, whereas LGLOB succeeded in finding the solution near $(1.3, 1)$ very frequently. When it did not succeed, it was because it reached an iterate where progress slowed dramatically, with continued improvement in P at each iteration, but typically only in the second decimal position. For example, for the starting point given in the last row of the table, this phenomenon occurred near the point $u = -15.688$, $v = 0.999$ (here $e^u \approx 0$ and u is close to a turning point of the cosine function). If a small random perturbation is made in this point and the LGLOB algorithm restarted (for example, at $u = -16$, $v = 1$), we found that it was then successful in finding the solution. A second instance is given in the next to last row of the table, and these two runs are therefore distinguished by S*. In constrast, when a similar perturbation strategy was attempted with DNEQNJ, it again failed (denoted by F* in the table). The reason is that the point where LGLOB slows down lies in a region where the Hessian matrix of P is ill-conditioned, but note that this region *does not contain a point of attraction* of the iteration sequence. In contrast, the region does contain a point of attraction for the sequence generated by DNEQNJ.

Obviously, a perturbation strategy is crude and would not be recommended as a general implementational technique. However, it does show that a restarting heuristic for LGLOB is on hand, which is not the case for DNEQNJ. The example highlights the need for careful attention to the case of ill-conditioning of the Hessian matrix of P (loss of rank in the Jacobian matrix of $\mathbf{h}$) and helps to explain why a general-purpose unconstrained optimization code for minimizing P and a general-purpose NLP code for solving (3.18) can both prove to be unsatisfactory.

3.3.4 LNC Algorithms

The foregoing algorithm demonstrates the viability of the underlying Lagrangian globalization approach. It has the advantage that the potential P defined by (3.21) can be used directly within a Newton–Cauchy minimization algorithm. But P has certain drawbacks, in particular, the following:

- The number of problem variables increases from n to $2n$;
- The derivatives of P are more expensive to compute than those of the sum-of-squares merit function F;
- The quantities c and α must be chosen artfully to exclude spurious critical points of P from a region containing a desired solution.

We now turn to more practical LNC algorithms that can alleviate these disadvantages.

For purposes of discussion, consider a cubic polynomial equation in a single variable $x \in R$ as follows:

$$h(x) = x^3 - 4x^2 + 4x + 1 = 0, \tag{3.34}$$

which has a unique root $x^* \approx -0.2055694304$. It also has a pair of complex conjugate roots, but as we have observed earlier, "equation-solving" in the context of this monograph means the task of finding a *real* root of a real equation or system of equations.

Now consider the sum-of-squares merit function $F(x)$ defined by (3.3) for the cubic polynomial. The point $\overline{x} = 2$ is a local minimizer of $F(x)$, but it is not a root of $h(x)$; i.e., the derivative of F vanishes at $\overline{x}$, but $h(\overline{x}) = 1$. Between the root x^* of h and the point $\overline{x}$, the function F has a "hump" that prevents a descent algorithm from reaching x^* when it is applied to F and initiated at or near the local minimizing point $\overline{x}$.

For the potential function defined by (3.21) and (3.34), let us choose $c = 10^5$ and $\alpha = 10^{-4}$. Then P has a minimizing point at

$$(x \approx -0.2055694, \ \lambda \approx 0.0360271)$$

and a spurious critical point at

$$(x \approx 1.3335347, \ \lambda \approx -10000.0).$$

Larger values of c and smaller values of α give spurious critical points with larger values of $|\lambda|$; i.e., spurious points are "pushed out to infinity" in accordance with Proposition 3.4 of Section 3.3.2.

When an equation-solving routine based on the merit function F, for example, the IMSL routine DNEQNJ of the previous section, is initiated at a point $x \geq 2$, it will converge falsely to the local minimizer of F at $x = 2$. In constrast, the potential P, as employed in the illustrative routine LGLOB of the previous section, introduces an *additional degree of freedom* via a Lagrange multiplier; i.e., its domain is a two-dimensional space of variables that permits escape from the unidimensional valley in the x-space. A minimization routine applied to P can go "around the hump" that blocks further progress toward the solution when F is employed. When the LGLOB routine[6] of Section 3.3.3 is applied to problem (3.34) and initiated from a point $x \geq 2$, it finds the solution x^* without becoming trapped by the local minimum of F at $x = 2$ or being drawn to the distant spurious critical point of P.

Consider again the more realistic example (3.7) of Watson, Billups, and Morgan [1987]. As noted there, the routine DNEQNJ produces the false solution (3.8). Suppose the LGLOB procedure is initiated at the point $\overline{\mathbf{x}}$

[6] A simplified version of this routine with $\mathbf{W}_k = \mathbf{I} \ \forall \ k$ is used in the experiments described in the present section.

given by (3.8) and its termination criterion is defined by

$$\|\mathbf{h}(\mathbf{x})\| \leq r\|\mathbf{h}(\overline{\mathbf{x}})\|, \tag{3.35}$$

with $0 < r < 1$, specifically, $r = 0.5$. Take $c = 10^4$ and $\alpha = 10^{-6}$, which are obtained from rough order-of-magnitude estimates of the quantities c_0 and b in Propositions 3.3 and 3.4 of Section 3.3.2, and choose the parameters σ and β as in Section 3.3.3. The procedure emerges after 32 iterations with the point

$$\tilde{\mathbf{x}} \approx (1.728, 1.673, 1.588, 1.479, 1.350, 1.206, 1.053, 0.895, 0.737, 0.585),$$

where $F(\tilde{\mathbf{x}}) \approx 1.845$. If the routine DNEQNJ is reinitiated at $\tilde{\mathbf{x}}$, it successfully finds the solution of the problem and terminates after 22 iterations with

$$\mathbf{x}^* \approx (2.620, 2.351, 1.980, 1.586, 1.227, 0.936, 0.717, 0.563, 0.461, 0.400)$$

and $F(\mathbf{x}^*) \approx 0$.

The Lagrangian globalization algorithm described in Nazareth and Qi [1996] builds on these ideas as illustrated by the foregoing numerical experiment. A traditional equation-solving algorithm based on the sum-of-squares merit function, which is often very effective in practice, is *interleaved* with the Lagrangian globalization approach to recover from cases where the former procedure fails. The LG procedure is terminated when the criterion (3.35) is satisfied. The traditional algorithm is terminated in the standard way; see Dennis and Schnabel [1983].

This combined approach is underpinned by the propositions of Section 3.3.2. In particular, Proposition 3.1 implies that an LG procedure can be initiated along a descent direction for P at any point where a procedure based on the sum-of-squares merit function F fails. This proposition also implies that $\mathbf{J}(\tilde{\mathbf{x}})^T\mathbf{h}(\tilde{\mathbf{x}}) \neq \mathbf{0}$; i.e., that the gradient vector of F is nonzero, at any critical point of P, say $(\tilde{\mathbf{x}}, \tilde{\lambda})$, where $\mathbf{h}(\tilde{\mathbf{x}}) \neq \mathbf{0}$.

Note that the quantity c_0, defined in Proposition 3.3, is *inversely* proportional to the size of the residual vector where the LG procedure is initiated. Thus, if the parameter c of P is chosen to satisfy $c \geq \frac{c_0}{r}$, where r is the quantity used in the termination criterion (3.35), then the potential P (employed at the initiating point of the LG phase) and the assertion of Proposition 3.3 remain valid for any subsequent iterate of the LG phase—denote it again by $\overline{\mathbf{x}}$—for which the termination criterion (3.35) is not yet satisfied.

The combined approach just described has *another important advantage.* Since P is *not* being minimized, it is possible to employ a more standard augmented Lagrangian with $\alpha = 0$, namely,

$$P(\mathbf{x}, \lambda; c) = L(\mathbf{x}, \lambda) + \tfrac{1}{2}c\|\mathbf{h}(\mathbf{x})\|^2, \tag{3.36}$$

where $c \in R_+$. In this case, every critical point of P in (3.36) is a solution of (3.1) and vice versa; i.e., there are no spurious critical points. (Proposition 3.4 of Section 3.3.2 can be satisfied trivially with $b = 0$.) The derivatives of (3.36) are less expensive to compute than the derivatives of the potential based on (3.21) with $\alpha > 0$.

There are numerous other possibilities for refinements of the Lagrangian globalization approach to nonlinear equation-solving, based on Lagrangian techniques surveyed in Bertsekas [1982]. These techniques can be additionally specialized or refined in order to take account of the special characteristics of (3.19) and (3.18). Note, in particular, the following:

- $\mathbf{h}(\mathbf{x}) = \mathbf{0}$ is a *square* system.
- There is freedom in the objective function $f(\mathbf{x})$, which can therefore be chosen (even varied) to advantage.
- The Newton–Cauchy techniques used to reduce P can be chosen suitably to match the needs of the problem. For example, conjugate gradient or limited-memory BFGS techniques can be used when n is large.

Considerable room remains for strengthening the propositions of Section 3.3.2, formulating viable LNC algorithms based on the Lagrangian globalization approach, proving their convergence, and developing effective implementations and software.

For some further discussion of the relative merits of the three main methods of this chapter, see Nazareth [1996].

3.4 Notes

Section 3.2: This material is derived from Nazareth [1995a].

Section 3.3: The material of this section is derived from Nazareth [1996a] and Nazareth and Qi [1996], where additional detail can be found.

Part II

Lessons from One Dimension

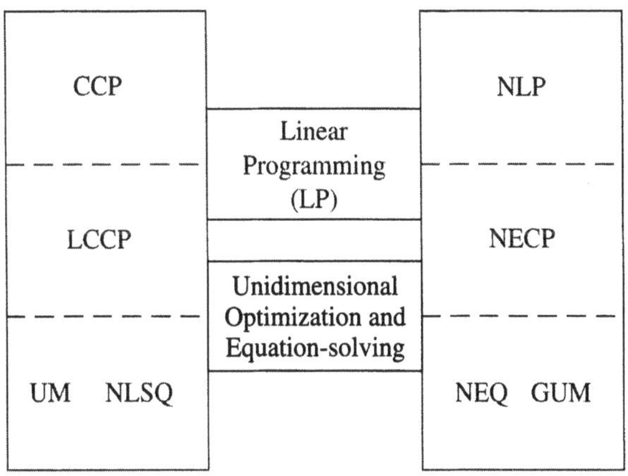

4

A Misleading Paradigm

The unidimensional problems considered in this chapter are as follows:

$$\text{minimize}_{x \in R} \qquad f(x) \tag{4.1}$$

and

$$\text{solve}_{x \in R} \qquad h(x) = 0. \tag{4.2}$$

Techniques for solving (4.1)–(4.2) are among the earliest discoveries of algorithmic differentiable programming. The subject remains surprisingly interesting from an algorithmic standpoint for several reasons:

- The *interplay* between algorithms for (4.1) and (4.2) for the unidimensional case has served as a *paradigm* for higher-dimensional optimization and equation-solving within the traditional development; see Figure 1.1.
- Unidimensional problems (4.1)–(4.2) occasionally arise *in practice*, and robust, efficient algorithms are needed to solve them.
- The analysis of the *rate of convergence* of unidimensional algorithms is a fascinating topic from a mathematical standpoint.
- Appropriate adaptations of unidimensional techniques form the backbone of *line-search algorithms* used within algorithms for solving higher-dimensional problems.
- Exploring *continuities* between univariate algorithms and their multidimensional counterparts can provide useful insights into the formulation of new algorithms.

We will consider the first three items in this chapter and the remaining two items in Chapters 5 and Chapter 6, respectively.

The following notation will be used for scalar quantities that arise in the discussion below; see also the comments on notation on page xvii. Let $g(x) = f'(x)$ denote the first derivative of f at $x \in R$ and let $J(x) = h'(x)$ denote the first derivative of h at x, i.e., its 1×1 symmetric, Jacobian matrix. Let x^* denote a solution of (4.1) or (4.2). We will assume that $J(x^*) = h'(x^*) \neq 0$ at a root x^* of (4.2), i.e., that it is a simple root. If g and h are identified with one another, so that $J(x^*) = f''(x^*) \neq 0$, then the corresponding assumption for f is that the critical point x^* is not also a point of inflection.

4.1 The Unidimensional Potential Function

The unidimensional case has been used as a paradigm for multidimensional differentiable programming in the traditional treatment (Figure 1.1). However, this paradigm is misleading. Unlike the multidimensional problem (3.1) of the previous chapter, the unidimensional function h in (4.2) always has a potential function, because its 1×1 Jacobian matrix is trivially symmetric. This potential function, say p, is found by integrating h or by appealing to (3.17) with $n = 1$, and p is unique up to an additive constant. The derivative of the potential function p is h, and its critical points are the roots of the system (4.2). Thus, solving a nonlinear equation in a single variable should be classified with solving higher-dimensional nonlinear systems with *symmetric* Jacobians rather than with solving *general* systems of nonlinear equations. And, in the same voice, algorithms for unidimensional optimization and algorithms for unidimensional equation-solving are much more closely interrelated than are their multidimensional counterparts.

For purposes of illustration, consider again example (3.34) of Chapter 3. The cubic polynomial $h(x) = x^3 - 4x^2 + 4x + 1$, $x \in R$, can be integrated, yielding a quartic polynomial function, say p, that is a potential function for h. Or equivalently, this quartic polynomial could be obtained from the construction (3.17). The only real-valued critical point of p is at $x^* \approx -0.2056$, and the second derivative of p at x^* is positive. Thus the root x^* of the cubic h could be found by a standard minimizing algorithm of descent applied to p.

For a general function h, the situation becomes more complicated. For example, an analytic expression for p may not be readily available, although *numerical integration* of (3.17) is always an option for obtaining the value $p(x)$ at any given point x, albeit an expensive one. Also, a root x^* of h could correspond to a (locally) maximizing point of p, so that a potential-function minimization approach must be applied to $-p$. Thus, in order to locate a root of h in general, one would need to minimize both p and $-p$. There is also the possibility, explicitly excluded by assumption in the introduction to this chapter, that x^* could correspond to a point of inflexion of p. For these reasons, the potential-function approach to root-finding may not always be useful from a *practical* standpoint. But nevertheless, it is very useful for purposes of conceptual discussion.

4.2 Globally Convergent Algorithms

In univariate equation-solving, there is no need to resort to potential functions of the types discussed in Section 4.1 or Section 3.3 in order to obtain a globally convergent method. The well-known *bisection algorithm*, which is globally convergent and simple to implement, is available for solving the nonlinear equation (4.2). Its only requirement is that a bracketing interval $[a, b]$ that satisfies $x^* \in [a, b]$ and $h(a)h(b) < 0$ be either on hand or able to be found, in order to initiate the algorithm.

For the minimization problem (4.1), the bisection algorithm can be used to solve $g(x) = 0$, where g denotes the derivative of f, provided that a two-point bracketing interval $[a, b]$ is available for which $g(a) < 0$ and $g(b) > 0$. (This implies that a minimizing point x^* lies in $[a, b]$.) If only function values are available and f is strictly *unimodal*, then *golden-section* search can be applied to a starting interval that brackets the (unique) minimizing point. For further detail on golden-section search and the closely related Fibonacci algorithm, see Luenberger [1984] and our discussion in Chapter 6.

For an arbitrary continuous function f in problem (4.1), the foregoing approach can be generalized by using a *three-point bracketing pattern*, i.e., a set of three points a, c, b such that $a < c < b$ and $f(a) \geq f(c) \leq f(b)$. A simple algorithm proceeds as follows:

1. Pick the larger of the two intervals $[a, c]$ and $[c, b]$. (If they are the same length, then pick either one.) Find its midpoint. For convenience, let us assume[1] that this interval is $[a, c]$ with midpoint $d = (a + c)/2$.
2. If $f(d) \geq f(c)$, then $a \leftarrow d$ and go to Step 4.
3. If $f(d) < f(c)$, then $b \leftarrow c$ and $c \leftarrow d$.
4. If $[a, b]$ is sufficiently small, then stop; else return to Step 1.

The procedure can be extended, if desired, to incorporate a golden-section ratio in place of choosing the midpoint.

Straightforward heuristic techniques can be employed to find an initiating interval for the above algorithms. An example, based on successive extrapolations, can be found in the next chapter, Section 5.4.

---

[1]Extending the description of the algorithm to the other case where d is the midpoint of $[c, b]$ is straightforward, and details are omitted.

4.3 Rapidly Convergent Local Algorithms

There is a wide variety of algorithms with fast *local* convergence that fit a *low-degree polynomial* to function and/or derivative information at the current iterate, say x_k, and a set of previous[2] iterates $x_{k-1}, x_{k-2}, \ldots$. The next iterate is then defined as a root of the polynomial in the equation-solving case and a stationary point of the polynomial in the optimization case. (In either case, the fitted polynomial is of degree low enough to ensure that the iterate is uniquely defined.) Some examples are as follows:

- For the equation-solving problem (4.2), *Newton's algorithm* fits a linear function, i.e., a polynomial of degree 1, to $h(x_k)$ and $h'(x_k)$ at the current iterate x_k, and defines x_{k+1} as its unique root. Equivalently, the algorithm fits a quadratic polynomial to the potential function information $p(x_k)$, $p'(x_k)$, and $p''(x_k)$ at x_k and defines x_{k+1} as the unique stationary point of this quadratic. Identify f with p and g with h to obtain Newton's algorithm for the optimization problem (4.1).

  Define the error e_k associated with the iterate x_k as $e_k = |x_k - x^*|$, where $|\cdot|$ denotes absolute value. Under the assumption that e_1 is sufficiently small, together with the problem assumptions stated in the introduction to this chapter, a simple convergence analysis reveals that *asymptotically*,[3] $e_{k+1} = Me_k^2$, where $M = h''(x^*)/(2h'(x^*))$. Thus Newton's algorithm for solving (4.2) converges locally at a quadratic rate. For the optimization case (4.1), replace h by g.

- For equation-solving (4.2), the *secant algorithm* fits a linear polynomial to the information $h(x_k)$ at the current iterate x_k and $h(x_{k-1})$ at the previous iterate x_{k-1}, and defines the next iterate to be its unique root. Alternatively, it can be formulated by fitting a quadratic polynomial to the potential function information $p(x_k)$, $p'(x_k)$ at x_k and $p'(x_{k-1})$ at x_{k-1}, or equivalently, to $p'(x_k)$ at x_k and $p(x_{k-1})$, $p'(x_{k-1})$ at x_{k-1}, and then choosing the next iterate as the stationary point of the quadratic. Again, identify f with p and g with h to obtain the secant algorithm for the optimization problem (4.1).

  A more delicate but also quite straightforward analysis shows that asymptotically, $e_{k+1} = Me_ke_{k-1}$, where M is the constant defined above. If we let $y_k = \log Me_k$, then $y_{k+1} = y_k + y_{k-1}$. This is the well-known Fibonacci difference equation. Its characteristic equation is $\tau^2 = \tau + 1$, whose roots are the golden-section ratio $\tau_1 = \left(1+\sqrt{5}\right)/2 \approx 1.618$ and $\tau_2 = \left(1-\sqrt{5}\right)/2$. Solutions of the Fibonacci difference equation are $y_k = A\tau_1^k + B\tau_2^k$, where A and B are arbitrary constants. This,

---

[2] Usually, this set has one or two members and possibly none.

[3] This can be interpreted to mean that equality holds in the limit. A more rigorous statement of the convergence results quoted in this chapter is unnecessary for purposes of the present discussion.

in turn, implies that $e_{k+1} = M^{(\tau_1-1)}e_k^{\tau_1}$ asymptotically. This is the beautiful result that the order of convergence of the secant algorithm is the golden-section ratio. See Luenberger [1984] for further detail.

- For the optimization problem (4.1), the *cubic-fit algorithm* fits a cubic polynomial to the information $(f(x_k), f'(x_k))$ at x_k and $(f(x_{k-1}), f'(x_{k-1}))$ at x_{k-1}. The next iterate x_{k+1} is the minimizing point of this cubic and is as follows:

$$x_{k+1} = x_k - (x_k - x_{k-1})\left[\frac{f'(x_k) + u_2 - u_1}{f'(x_k) - f'(x_{k-1}) + 2u_2}\right], \tag{4.3}$$

where

$$u_1 = f'(x_{k-1}) + f'(x_k) - 3\frac{f(x_{k-1}) - f(x_k)}{x_{k-1} - x_k},$$

$$u_2 = \left[u_1^2 - f'(x_{k-1})f'(x_k)\right]^{1/2}.$$

These expressions are employed in an implementation in the next chapter; see Section 5.4.

The errors at successive iterates are related asymptotically by $e_{k+1} = Me_ke_{k-1}^2$, where M is defined in the first item above on Newton's algorithm. Let $y_k = \log\left(\sqrt{M}e_k\right)$. Then one obtains the difference equation $y_{k+1} = y_k + 2y_{k-1}$. Its characteristic equation is $\tau^2 = \tau + 2$, whose larger root is 2. This, in turn, implies that the order of convergence of the cubic-fit algorithm is 2.

The cubic-fit algorithm can be applied to the equation-solving problem (4.2) by identifying the potential p with f in the foregoing development. Note that the expression for the next iterate derived from the cubic fit would require the values of the potential at the current and previous iterates. This information is available numerically as discussed in Section 4.1, but the expense of computing it may make the approach impractical.

- For the optimization problem (4.1), the *quadratic-fit algorithm* fits a quadratic polynomial $q(x)$ to the information $f(x_k)$ at x_k, $f(x_{k-1})$ at x_{k-1}, and $f(x_{k-2})$ at x_{k-2}. This quadratic is given by the standard Lagrange interpolation formula as follows:

$$q(x) = \sum_{i=k-2}^{k} f(x_i)\frac{\prod_{j\neq i}(x - x_j)}{\prod_{j\neq i}(x_i - x_j)}. \tag{4.4}$$

The next iterate x_{k+1} is the stationary point of q. Errors at successive iterates are related, in the limit, by $e_{k+1} = Me_{k-1}e_{k-2}$, where M is defined above. With the definition y_k as in the secant method, one obtains the difference equation $y_{k+1} = y_{k-1} + y_{k-2}$, with characteristic equation $\tau^3 = \tau + 1$ and largest root $\tau_1 \approx 1.3$. Again, as above, τ_1 determines the order of convergence of the algorithm.

A quadratic-fit algorithm can be applied to the potential function p associated with the equation-solving problem (4.2); see the analogous discussion for the cubic-fit algorithm.

A detailed discussion of the foregoing algorithms and associated convergence analysis can be found in Luenberger [1984].

There are numerous other unidimensional algorithms based on higher-order polynomial fitting. A unified discussion is given in the classic monograph of Traub [1964].

4.4 Practical Hybrid Algorithms

A viable algorithm must be globally convergent as in Section 4.2 and exhibit fast local convergence as in Section 4.3. Consider, for example, the well-known variant of the secant algorithm for equation-solving where the iterates to which the linear polynomial is fitted are chosen so that they always bracket a root; i.e., they are not necessarily the two most recent iterates. This strategy ensures global convergence, but examples can easily be constructed where the starting point is retained for all iterations. The rate of convergence of the resulting algorithm is then only linear. Similarly, for the optimization problem (4.1), a quadratic fit can be made at points that are constrained to form a three-point bracketing pattern, or a cubic fit can be made using a two-point pattern based on derivatives; see the analogous discussion on bisection in Section 4.2—yielding algorithmic variants with weakened rate-of-convergence properties.

Practical algorithms are similar to the above, but they employ a more *artfully constructed hybrid* of the globally convergent and rapid locally convergent techniques of Sections 4.2 and 4.3. Their objective is global convergence, superlinear local convergence, and numerical stablity. For details of implementation, consult the classic monograph of Brent [1973].

4.5 Summary

The interplay between algorithms of univariate optimization and algorithms of equation-solving should *not* be used as a paradigm for their multidimensional counterparts. Rather, the subject of univariate, differentiable programming provides a bridge between the unconstrained minimization and nonlinear equation-solving areas, as summarized in Figure 1.2.

From any practical standpoint, (4.1) and (4.2) are very insignificant *problems* of science and engineering. However, their main solution algorithms are *mathematical objects* of intrinsic interest, and thus well worthy of study. They lend themselves to a beautiful rate-of-convergence analysis, and the development of numerically sound, practically efficient hybrids is also a challenging task. As noted earlier, the classic monographs in these two areas are, respectively, Traub [1964] and Brent [1973], and both are of continued relevance today.

More recent algorithmic developments can be studied in a similar vein, for example, univariate algorithms based on *conic* models proposed for optimization by Davidon [1980] and their counterpart *rational* models for equation-solving.

The complexity analysis of univariate programming algorithms has also been a focus of attention in recent years; see, for example, Nemirovsky and Yudin [1983] and the monograph of Sikorski [2001].

5

CG and the Line Search

We now consider the use of univariate techniques of the previous chapter within the context of algorithms for solving optimization problems in several variables, specifically, three key enhancements of Cauchy's method for unconstrained minimization that were published, coincidentally and independently of one another, in the same year, 1964. These three algorithms are as follows:

- the nonlinear conjugate-gradient (CG) algorithm of Fletcher and Reeves [1964];
- the method of parallel tangents, or PARTAN algorithm, of Shah, Buehler, and Kempthorne [1964];
- the heavy-ball algorithm of Polyak [1964].

The Fletcher–Reeves algorithm was directly motivated by the *linear* CG algorithm of Hestenes and Stiefel [1952]. The other two algorithms, PARTAN and heavy-ball, were developed from quite different vantage points, but they were also shown later to have a close relationship with the Hestenes–Stiefel algorithm.

Fletcher and Reeves [1964] employed a line-search technique, which was based on an extrapolation/cubic interpolation strategy adapted from Davidon [1959], in order to develop an *implementable* version of their algorithm. This choice made all the difference. The key role of the line search in the good fortune of the Fletcher–Reeves CG algorithm, in relation to its other two 1964 competitors, is the topic of the present chapter; see, in particular, Section 5.3. During the course of this discussion we also elaborate on the important distinction between *conceptual* and *implementable* algorithms (Polak [1971]), which was introduced in Chapter 3.

A Fortran program for a version of the extrapolation/cubic interpolation line search mentioned above is provided in Section 5.4—a detailed embodiment of the univariate techniques introduced in Chapter 4—and it is used for purposes of implementation in the present chapter and in Chapters 9 and 14.

The future prospects of nonlinear CG-related algorithms are considered in the concluding Section 5.5, and in particular, a new two-parameter family of nonlinear CG algorithms is formulated. Some numerical experiments with it are also reported. This CG family will be used again later, in Chapter 14, to provide a detailed illustration of a new population-based algorithmic paradigm for optimization and equation-solving.

5.1 The Linear CG Algorithm

The conjugate-gradient algorithm was originally proposed by Hestenes and Stiefel [1952] for solving a system of linear equations $\mathbf{Ax} = \mathbf{b}$ for the unknown vector $\mathbf{x} \in R^n$, where $\mathbf{A}$ is a positive definite symmetric matrix and $\mathbf{b}$ is a given vector. Usually, n is large. For background information, see, for example, Golub and O'Leary [1989] and O'Leary [1996]. The mapping $\mathbf{Ax} - \mathbf{b}$ has a *potential function*, say p, which is given by expression (3.17), namely,

$$p(\mathbf{x}) = -\mathbf{b}^T\mathbf{x} + \frac{1}{2}\mathbf{x}^T\mathbf{A}\mathbf{x}. \tag{5.1}$$

Therefore, solving $\mathbf{Ax} = \mathbf{b}$, where $\mathbf{A}$ is positive definite and symmetric, is mathematically equivalent to finding the unique minimizing point of a strictly convex quadratic function.

A basic version of the Hestenes–Stiefel algorithm is as follows:

Initialization:

$\mathbf{x}_1$ = arbitrary;

$\mathbf{r}_1$ = residual of the linear system (gradient of the potential) at $\mathbf{x}_1$;

$\mathbf{d}_1 = -\mathbf{r}_1$;

Iteration k:

1. $\mathbf{x}_{k+1}$ = unique minimizing point of the potential p on the half-line through $\mathbf{x}_k$ along the direction $\mathbf{d}_k$;
2. $\mathbf{r}_{k+1}$ = residual of the linear system (gradient of the potential) at $\mathbf{x}_{k+1}$;

3. $\beta_k = \mathbf{r}_{k+1}^T \mathbf{y}_k / \mathbf{d}_k^T \mathbf{y}_k$ where $\mathbf{y}_k = \mathbf{r}_{k+1} - \mathbf{r}_k$;
4. $\mathbf{d}_{k+1} = -\mathbf{r}_{k+1} + \beta_k \mathbf{d}_k$.

The matrix $\mathbf{A}$ is provided exogenously. The residual $\mathbf{r}_{k+1}$, at step 2, is computed as $\mathbf{A}\mathbf{x}_{k+1} - \mathbf{b}$, or else it is obtained by updating $\mathbf{r}_k$. The direction $\mathbf{d}_k$ is always a descent direction for p at $\mathbf{x}_k$. At step 1, the minimizing point is computed as follows:

$$\alpha_k = -\frac{\mathbf{r}_k^T \mathbf{d}_k}{\mathbf{d}_k^T \mathbf{A} \mathbf{d}_k}, \quad \mathbf{x}_{k+1} = \mathbf{x}_k + \alpha_k \mathbf{d}_k. \tag{5.2}$$

In step 4, observe that $\mathbf{d}_{k+1}^T \mathbf{y}_k = 0 = \mathbf{d}_{k+1}^T \mathbf{A} \mathbf{d}_k$; i.e., the directions $\mathbf{d}_{k+1}$ and $\mathbf{d}_k$ are conjugate. Furthermore, it is easy to show that $\mathbf{d}_{k+1}$ is conjugate to *all* previous search directions and that the algorithm terminates in at most n steps in exact arithmetic.

The following relations can also be easily proved:

$$\mathbf{r}_{k+1}^T \mathbf{r}_k = 0, \quad \mathbf{r}_{k+1}^T \mathbf{r}_{k+1} = \mathbf{r}_{k+1}^T \mathbf{y}_k, \quad \mathbf{r}_k^T \mathbf{r}_k = \mathbf{d}_k^T \mathbf{y}_k. \tag{5.3}$$

They imply that there are several alternative expressions for the quantities in the above algorithm, in particular, for β_k (see also Section 5.5.1). For example, $\beta_k = \mathbf{r}_{k+1}^T \mathbf{r}_{k+1} / \mathbf{r}_k^T \mathbf{r}_k$ and hence $\beta_k > 0$, because the residual vanishes only at the solution. There are numerous variants of the basic CG algorithm that are designed to enhance convergence and improve algorithm stability, through the use of problem preconditioning (transformation of variables), and/or to solve related problems of computational linear algebra. A contextual overview and further references can be found in Barrett et al. [1993].

When the focus is on minimizing the potential function p rather than on solving the system of positive definite symmetric linear equations, the residual at $\mathbf{x}_{k+1}$ is given its alternative interpretation and representation as the gradient vector $\mathbf{g}_{k+1}$ of p at $\mathbf{x}_{k+1}$, and *this gradient is assumed to be provided exogenously.* The minimizing point at step 1 is computed, alternatively, as follows:

$$\alpha_k = -\frac{\mathbf{g}_k^T (\bar{\mathbf{x}}_k - \mathbf{x}_k)}{\mathbf{d}_k^T (\bar{\mathbf{g}}_k - \mathbf{g}_k)}, \tag{5.4}$$

where $\bar{\mathbf{x}}_k$ is *any* point on the ray through $\mathbf{x}_k$ in the direction $\mathbf{d}_k$, and $\bar{\mathbf{g}}_k$ is its corresponding gradient vector. The foregoing expression for α_k is derived from the linear systems version (5.2) and the relation $\mathbf{A}(\bar{\mathbf{x}}_k - \mathbf{x}_k) = \bar{\mathbf{g}}_k - \mathbf{g}_k$. Again, $\mathbf{x}_{k+1} = \mathbf{x}_k + \alpha_k \mathbf{d}_k$.

The resulting optimization algorithm will be called the *CG standard* for minimizing p. It will provide an important guideline for defining nonlinear CG algorithms in the subsequent discussion.

5.2 Nonlinear CG-Related Algorithms

5.2.1 The Nonlinear CG Algorithm

The *nonlinear CG method* addresses the unconstrained minimization problem (1.2), usually when n is large or computer storage is at a premium. As discussed in Chapter 2, optimization techniques for this class of problems, which are inherently iterative in nature, form a direction of descent at the current iterate (approximating the solution), and move along this direction to a new iterate with an improved function value. The nonlinear CG algorithm was a marked enhancement of the classical steepest-descent method of Cauchy. Like steepest descent, it is storage-efficient and requires only a few n-vectors of computer storage beyond that needed to specify the problem itself.

A basic conceptual version of the original Fletcher–Reeves CG algorithm is as follows:

Initialization:

$\mathbf{x}_1$ = arbitrary;

$\mathbf{g}_1$ = gradient of f at $\mathbf{x}_1$;

$\mathbf{d}_1 = -\mathbf{g}_1$;

Iteration k:

1. $\mathbf{x}_{k+1}$ = global minimizing point on ray through $\mathbf{x}_k$ along direction $\mathbf{d}_k$;
2. $\mathbf{g}_{k+1}$ = gradient of f at $\mathbf{x}_{k+1}$;
3. $\beta_k = \mathbf{g}_{k+1}^T\mathbf{g}_{k+1}/\mathbf{g}_k^T\mathbf{g}_k$;
4. $\mathbf{d}_{k+1} = -\mathbf{g}_{k+1} + \beta_k\mathbf{d}_k$.

Note that this algorithm is closely patterned after the CG standard and replicates it precisely on a strictly convex quadratic (see also the remarks following (5.3)). *This property characterizes a nonlinear CG algorithm.*

The choice of $\mathbf{x}_{k+1}$ in step 1 implies that $\mathbf{g}_{k+1}^T\mathbf{d}_k = 0$ and, in step 4, $\mathbf{g}_{k+1}^T\mathbf{d}_{k+1} < 0$. Thus $\mathbf{d}_{k+1}$ is always a direction of descent. The basic Fletcher–Reeves algorithm defined above, *without restarts*, is globally convergent (Al-Baali [1985]). See also our subsequent discussion in Section 5.5.1. A common strategy is to restart the algorithm periodically along the negative gradient direction. In this case global convergence is a consequence of a more general, well-known result on the convergence of restarted descent algorithms; see, for example, Luenberger [1984].

5.2.2 The PARTAN Algorithm

A basic conceptual version of the parallel-tangents algorithm of Shah, Buehler, and Kempthorne [1964] is as follows:

Initialization and iteration $k = 1$:

$\mathbf{x}_1$ = arbitrary;

$\mathbf{g}_1$ = gradient of f at $\mathbf{x}_1$;

$\mathbf{x}_2$ = global minimizing point on the ray through $\mathbf{x}_1$ along direction $-\mathbf{g}_1$;

$\mathbf{g}_2$ = gradient of f at $\mathbf{x}_2$;

Iteration $k \geq 2$:

1a. $\mathbf{z}_k$ = global minimizing point on ray through $\mathbf{x}_k$ along direction $-\mathbf{g}_k$;

1b. $\mathbf{x}_{k+1}$ = global minimizing point on the line joining $\mathbf{x}_{k-1}$ and $\mathbf{z}_k$;

2. $\mathbf{g}_{k+1}$ = gradient of f at $\mathbf{x}_{k+1}$;

Each point $\mathbf{x}_k$ is called a major iterate, and each $\mathbf{z}_k$ is called a minor iterate. The PARTAN algorithm is globally convergent, because it is at least as good as steepest descent at each major iterate $\mathbf{x}_k$. (For this reason, restarts are not needed.) The algorithm can be shown to cover the path of the CG standard on a strictly convex quadratic potential p, because the major iterates $\mathbf{x}_k$ and the directions $\mathbf{d}_k$ defined by successive major iterates, namely, $\mathbf{d}_k = \mathbf{x}_{k+1} - \mathbf{x}_k$, $k = 1, \ldots, n$, are the same as the iterates and directions[1] of the CG standard. For details, see Luenberger [1984].

The PARTAN algorithm derives its name from "parallel tangents," a special geometric property of the tangents to the contours of a quadratic function. This original motivation has been supplanted by the CG connection, which was recognized subsequently. More recently, Nesterov [1983] has given an important result concerning the *global rate of convergence* on convex functions of an algorithm very similar to PARTAN.

5.2.3 *The Heavy-Ball Algorithm*

The heavy-ball algorithm of Polyak [1964] is defined by the recurrence relation

$$\mathbf{x}_{k+1} = \mathbf{x}_k - \alpha_k \mathbf{g}_k + \beta_k(\mathbf{x}_k - \mathbf{x}_{k-1}), \ k \geq 1, \tag{5.5}$$

where $\mathbf{x}_1$ is given, $\mathbf{x}_0 \equiv \mathbf{0}$, $\alpha_k > 0$ and $\beta_k \geq 0$ are scalar quantities discussed below, and $\beta_1 \equiv 0$. The method derives its name from a second-order differential equation describing the smooth trajectory of a body ("the heavy ball") in a potential field associated with the function f under a force of friction. The iterative process (5.5) is the difference analogue of this differential equation, with the second term representing an inertial vector designed to speed convergence. For details, see Polyak [1987].

When α_k is a constant positive step size, say $\alpha > 0$, and β_k is a scalar typically chosen in the range $0 \leq \beta_k < 1$, then (5.5) defines a simple multistep algorithm (more specifically, a two-step algorithm). For a discussion of

[1] Up to a positive scalar multiple.

how to make good, implementable choices for these parameters, see Polyak [1987], Bertsekas [1999].

A second, *conceptual* approach is to choose the parameters by solving the two-dimensional optimization problem as follows:

$$\{\alpha_k, \beta_k\} = \mathrm{argmin}_{\alpha,\beta} f(\mathbf{x}_k - \alpha \mathbf{g}_k + \beta(\mathbf{x}_k - \mathbf{x}_{k-1})). \qquad (5.6)$$

When the algorithm is applied to a strictly convex quadratic function, the problem (5.6) has an explicit solution. Furthermore, it is easy to show that the resulting algorithm generates the same iterates as the CG standard on a strictly convex quadratic function. For details, see again Polyak [1987].

5.3 The Key Role of the Line Search

The linear CG algorithm of Section 5.1 can be implemented directly using (5.2) or (5.4). In contrast, the nonlinear CG-related algorithms of Section 5.2 are called *conceptual*, because they use procedures that cannot generally be implemented *in a finite number of operations*; i.e., they employ exact minimizations over a ray or a two-dimensional space. The important distinction between *conceptual* and *implementable* algorithms and an associated hierarchical formulation of algorithms are due to Polak [1971]. These ideas were introduced in Sections 3.2.2 and 3.2.3, and they will also be explored further, within the context of models of computation, in Chapter 15.

In order to develop implementable versions of the algorithms of Section 5.2, one can substitute simple step-length procedures to move from the current iterate to the next. The PARTAN and heavy-ball algorithms were originally formulated in this vein. In contrast, the nonlinear CG algorithm implemented by Fletcher and Reeves [1964] used the extrapolation/cubic interpolation line-search technique of Davidon [1959] and a termination criterion that resembled the subsequently formulated, and now standard, strong Wolfe conditions (Wolfe [1969]). When applied to a strictly convex quadratic function, such a line search procedure can immediately locate the minimizing point along the direction of search, once it has gathered the requisite information to make an exact fit. The effectiveness of the Davidon [1959] line search played a considerable role in the success of the Fletcher–Reeves algorithm, and its use within either of the other two 1964 algorithms could well have propelled them into the limelight at that time.

In contrast to the nonlinear CG and PARTAN algorithms, which use line searches, it is worth noting that the heavy-ball algorithm is best suited to a *two-dimensional* search procedure. A technique based on quadratic approximation that uses only function values and is efficient in spaces of low dimensions has been developed recently by Powell [2000]. An interesting unexplored avenue of research would be the adaptation of this so-called

UObyQA algorithm[2] in conjunction with suitable termination criteria—the analogue of the Wolfe criteria—to meet the two-dimensional search requirements of the heavy-ball, CG-related algorithm. Since a conceptual two-dimensional search procedure subsumes an exact line search along the negative gradient direction at the current iterate, the proof of convergence of the conceptual algorithm on general functions is considerably simplified.

5.4 A Line-Search Fortran Routine

The following is an example of a line-search procedure written in Fortran that would be suitable for use within a CG-related algorithm:

```
      SUBROUTINE LNESRCH(N,X0,F0,G0,FP0,D0,ALPHA,FCNG,
     *                   TOL,STEPMX,ACC,EXTBND,
     *                   MAXFCN,NFCALL,IERR,
     *                   X,F,G)
      IMPLICIT REAL*8(A-H,O-Z)
      DOUBLE PRECISION X0(N),G0(N),D0(N),X(N),G(N)
      EXTERNAL FCNG
C
C*****
C
C   The problem is of dimension N. The current iterate is X0
C   with associated function value F0 and gradient G0. The search
C   direction is D0 and the directional derivative along it is FP0.
C   ALPHA defines the initial step along D0, for example, 1.0.
C   All of these quantities are input to the subroutine. FCNG is
C   an externally supplied subroutine with calling sequence
C   FCNG(N,x,f,g), which returns the function value f and gradient
C   vector g at any given point x.
C
C   TOLX, STEPMX, ACC, and EXTBND are tolerances that are used in
C   the line search subroutine (see the relevant statements below).
C   In particular, 0 < ACC < 1 determines the accuracy that must be
C   achieved on exit. The new iterate is returned in X0 with
C   associated information in F0 and G0.
C
C   No more than MAXFCN calls to the subroutine FCNG are allowed.
C   The actual number of calls are returned in NFCALL, and IERR is
C   a flag that specifies error information (if nonzero on exit).
C
C   X, F, and G are work space for the line search routine.
C
C   The cubic interpolation formulae used below are given in
```

[2]Unconstrained Optimization by Quadratic Approximation algorithm.

```
C   Chapter 4. See, in particular, the third item of Section 4.3
C   and expression (4.3). The LNESRCH Fortran code is derived from
C   Hillstrom [1976a], which is, in turn, derived from a routine
C   in the Harwell library.
C
C*****
C
      IERR=0
      NFCALL=0
C
  100 TOT=0.0D0
      CDERIV=FP0
      PDERIV=FP0
      FTEST=F0
C
  105 IF (ALPHA.LE.TOL) THEN
         IERR=2
         GOTO 150
      ENDIF
C
C   begin the line search by incrementing the solution vector and
C   calculating the function and gradient at the incremented vector.
C
      DO 108 I=1,N
         X(I)=X0(I)
         G(I)=G0(I)
         X0(I)=X0(I) + ALPHA*D0(I)
  108 CONTINUE
      F=FTEST
C
      CALL FCNG(N,X0,FTEST,G0)
C
      NFCALL=NFCALL+1
      IF (NFCALL.GT.MAXFCN) THEN
         IERR=1
         RETURN
      ENDIF
C
C   compute the directional derivative
c
      FP0=0.0D0
      DO 110 I=1,N
         FP0=FP0 + G0(I)*D0(I)
  110 CONTINUE
C
C   test whether the function has decreased
C
      IF (FTEST.GE.F) GOTO 120
C
```

```
C   sufficient progress and convergence criterion for exit
C
    IF (DABS(FPO/PDERIV).LE.ACC) GOTO 140
C
  116 CONTINUE
C
C   test for extrapolation or interpolation
C
    IF (FPO.GT. 0.0D0) GOTO 120
C
C   revise ALPHA by extrapolation, limiting the length of
C   the multiplier by EXTBND
C
    TOT=TOT+ALPHA
    IF (TOT.GT.STEPMX) GOTO 145
    TEMP=EXTBND
    IF (CDERIV.LT.FPO) TEMP=FPO/(CDERIV-FPO)
    IF (TEMP.GT.EXTBND) TEMP=EXTBND
    F=FTEST
    CDERIV=FPO
    ALPHA=ALPHA*TEMP
    GOTO 105
C
C   reset solution vector and revise ALPHA by cubic interpolation
C

  120 CONTINUE
    DO 130 I=1,N
       XO(I)=X(I)
       GO(I)=G(I)
  130 CONTINUE

C
    TEMP=3.0D0*(F-FTEST)/ALPHA + FPO + CDERIV
    WT=DMAX1(DABS(TEMP),DABS(FPO),DABS(CDERIV))
    WW=TEMP/WT
    WW=WW*WW - (CDERIV/WT)*(FPO/WT)
    IF (WW.LT.0.0D0) WW=0.0D0
    WW=DSQRT(WW)*WT
    TEMP=1 - (FPO + WW - TEMP)/(2*WW + FPO - CDERIV)
    ALPHA=ALPHA*TEMP
    FTEST=F
    GOTO 105
C
C   ALPHA is accepted
C
  140 FO=FTEST
  145 ALPHA=TOT+ALPHA
C
```

```
  150 CONTINUE
      RETURN
C
      END
```

The termination criteria used by the above line search routine LNESRCH are closely related, but not equivalent, to the strong Wolfe conditions (Wolfe [1969]). For a discussion of the latter, see, for example, Nazareth [1994b].

LNESRCH is used in numerical experiments described in the next section, and also in Chapters 9 and 14 of this monograph. In the terminology of Section 15.5.2, LNESRCH is a level-2 routine. More sophisticated level-3 line search routines are available that conform more closely to the standards of high-quality mathematical software as described in Section 15.5.3; see, for example, Moré and Thuente [1994].

5.5 CG: Whither or Wither?

5.5.1 *Other Nonlinear CG Algorithms*

There are many alternative expressions for the quantity β_k in the nonlinear CG algorithm of Section 5.2.1. The four leading contenders are the choices proposed by Hestenes and Stiefel [1952], Fletcher and Reeves [1964], Polyak [1969] and Polak and Ribière [1969], and Dai and Yuan [1999]. We will henceforth identify them by the abbreviations HS, FR, PPR, and DY, respectively. They define β_k as follows:

$$\text{HS: } \beta_k = \frac{\mathbf{g}_{k+1}^T \mathbf{y}_k}{\mathbf{d}_k^T \mathbf{y}_k}; \qquad \text{PPR: } \beta_k = \frac{\mathbf{g}_{k+1}^T \mathbf{y}_k}{\mathbf{g}_k^T \mathbf{g}_k}; \tag{5.7}$$

$$\text{FR: } \beta_k = \frac{\mathbf{g}_{k+1}^T \mathbf{g}_{k+1}}{\mathbf{g}_k^T \mathbf{g}_k}; \qquad \text{DY: } \beta_k = \frac{\mathbf{g}_{k+1}^T \mathbf{g}_{k+1}}{\mathbf{d}_k^T \mathbf{y}_k}; \tag{5.8}$$

where $\mathbf{y}_k = \mathbf{g}_{k+1} - \mathbf{g}_k$ is the gradient change that corresponds to the step $\mathbf{s}_k = \mathbf{x}_{k+1} - \mathbf{x}_k$.

When the objective function is a strict convex quadratic and an exact line search is used, the following relations are a direct consequence of (5.3):

$$\mathbf{g}_{k+1}^T \mathbf{g}_{k+1} = \mathbf{g}_{k+1}^T \mathbf{y}_k, \qquad \mathbf{g}_k^T \mathbf{g}_k = \mathbf{d}_k^T \mathbf{y}_k; \tag{5.9}$$

and then the values of the scalar β_k are *identical* for all four choices (5.7)–(5.8), and each of the associated algorithms becomes the CG standard. Termination occurs in at most n steps.

Moreover, for an arbitrary smooth objective function and an exact line search, the second of the two relations (5.9) still holds, but not necessarily the first, and thus in this more general setting,

$$\text{HS} \equiv \text{PPR}; \qquad \text{FR} \equiv \text{DY}.$$

A nonlinear CG algorithm with an exact line search, i.e., a *conceptual* nonlinear CG algorithm, always produces a direction of descent. The conceptual FR/DY algorithm without restarts is known to be globally convergent (Al-Baali [1985]), and the conceptual PPR/HS algorithm without restarts can cycle (Powell [1984]).

When the algorithms are made *implementable*, through the use of an *inexact* line search, and they are applied to nonquadratic functions, then the different choices for β_k in (5.7)–(5.8) result in four distinct nonlinear CG algorithms that can exhibit behaviors very different from one another. Suitable inexact line search termination conditions are, for example, the strong and weak Wolfe conditions; Wolfe [1969]. Their use in conjunction with the four different choices of β_k has been extensively studied, from both theoretical and computational standpoints.

A nonlinear CG algorithm can be guaranteed to produce a direction of descent $\mathbf{d}_{k+1}$, provided the line search is sufficiently accurate. The descent condition on $\mathbf{d}_{k+1}$ is

$$\mathbf{d}_{k+1}^T \mathbf{g}_{k+1} = -\mathbf{g}_{k+1}^T \mathbf{g}_{k+1} + \beta_k \mathbf{g}_{k+1}^T \mathbf{d}_k < 0.$$

In some algorithms, this condition follows as a consequence of the strong or weak Wolfe exit conditions within the line search. In other algorithms, the foregoing descent condition may have to be imposed as an additional, exogenous requirement on the line search. A good overview of these considerations can be found in the papers of Dai and Yuan, [1996], [1999], who have made important recent contributions to the study of nonlinear CG algorithms. See also other references cited in these articles.

Theoretical studies, such as the foregoing, often assume that no restarts are used. With periodic restarting, nonlinear CG algorithms whose search directions are directions of descent at each iteration are well known to be globally convergent; see, for example, Luenberger [1984].

Many other CG variants have been introduced, for example, the following:

1. $\beta_k = \max\{0, \mathbf{g}_{k+1}^T \mathbf{y}_k / \mathbf{g}_k^T \mathbf{g}_k\}$ (nonnegative PPR or PPR$^+$);
2. $\beta_k = -\mathbf{g}_{k+1}^T \mathbf{g}_{k+1} / \mathbf{d}_k^T \mathbf{g}_k$ (conjugate descent);
3. $\beta_k = -\mathbf{g}_{k+1}^T \mathbf{y}_k / \mathbf{d}_k^T \mathbf{g}_k$.

The last two expressions can result in large values for β_k when the directional derivative along the prior search direction $\mathbf{d}_k$ is small in magnitude, but the associated gradient vectors are not small. The next direction $\mathbf{d}_{k+1}$ will then be biased toward the previous, unsatisfactory, direction $\mathbf{d}_k$. Thus, these last two choices are unlikely to be of advantage in the nonlinear CG setting.

As a practical matter, the nonlinear CG algorithm of Powell [1977], which is based on the PPR choice for β_k and a suitable restarting strategy, has emerged as one of the most efficient, and it has provided the foundation for current nonlinear CG software. If a small step is taken from $\mathbf{x}_k$ to $\mathbf{x}_{k+1}$, but the associated gradient vectors are not small, then the PPR choice for β_k has the advantage of producing an automatic bias toward the direction of steepest descent; i.e., it has a built-in self-restarting mechanism. This is in marked contrast to many of the other CG choices mentioned above and is often cited as a reason for making the PPR or PPR$^+$ choices in practice.

5.5.2 A New Two-Parameter CG Family

The following generalization of (5.7)–(5.8) yields a new two-parameter family of nonlinear CG algorithms:

$$\beta_k = \frac{\lambda_k(\mathbf{g}_{k+1}^T\mathbf{g}_{k+1}) + (1-\lambda_k)(\mathbf{g}_{k+1}^T\mathbf{y}_k)}{\mu_k(\mathbf{g}_k^T\mathbf{g}_k) + (1-\mu_k)(\mathbf{d}_k^T\mathbf{y}_k)}, \tag{5.10}$$

where $\lambda_k \in [0,1]$ and $\mu_k \in [0,1]$. For any choice of λ_k and μ_k in these ranges, the associated algorithm reduces to the CG standard when f is quadratic and line searches are exact. This follows directly from (5.9). If the scalars λ_k and μ_k take only their extreme values, 0 and 1, then one obtains the four possible combinations (5.7)–(5.8). The denominator in (5.10) is positive for all $\mu_k \in [0,1]$, because the standard exit conditions on the line search ensure that $\mathbf{d}_k^T\mathbf{y}_k > 0$. The parameter λ_k can be chosen so that the numerator is also positive, if desired.

The subfamily of (5.10) with $\lambda_k \equiv 1$ was proposed independently by Dai and Yuan [1998] and studied in some detail by these authors. Note also that the parameters λ_k and μ_k could be chosen outside the range $[0,1]$ or defined to be scalar-valued functions, as with β_k in (5.7)–(5.8), but we will not pursue these extensions further here.

Henceforth, if the parameters λ_k and μ_k are held fixed within a nonlinear CG algorithm based on (5.10), i.e., they do not vary with k, then we will denote them more simply by λ and μ.

5.5.3 Illustration of Performance

Figure 5.1 illustrates the enormous variability in performance of different members of the new nonlinear CG family. The CG family is implemented along the lines of the nonlinear CG algorithm of Section 5.2.1 with β_k defined by (5.10), $\mu \equiv 1$, and λ ranging from 0 to 1 in increments of 0.05. The conceptual algorithm of that section is made implementable via the line search routine LNESRCH of Section 5.4 with a stringent choice for the termination parameter ACC $= 0.1$; i.e., a fairly accurate line search is performed at each iteration. No restarts are used. The implementation is exercised on Rosenbrock's function, as defined by expression (2.1), and uses its standard starting point $(-1.2,\ 1)$.

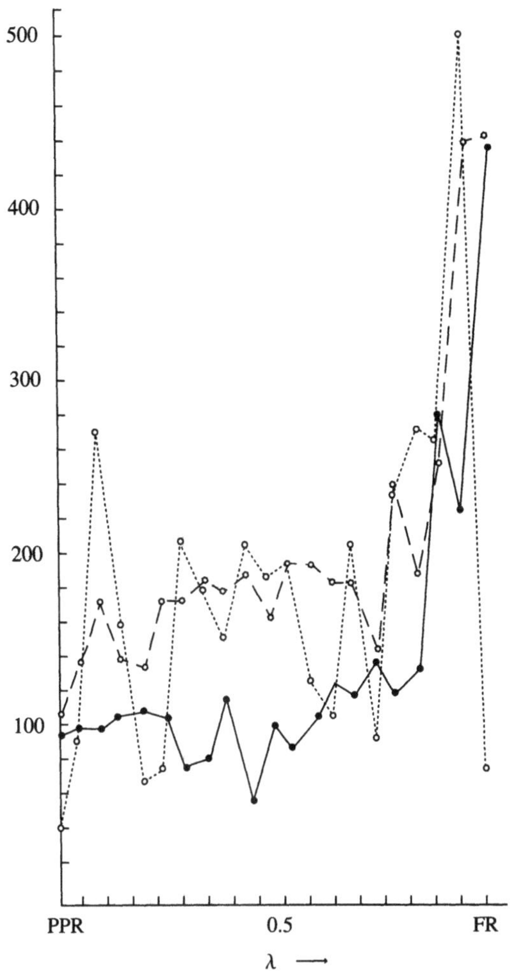

FIGURE 5.1 Performance of the CG family for three starting points.

The number of calls to the function/gradient evaluation routine FCNG is recorded for each run as defined by a particular choice of λ. These counts are then plotted for the different values of λ and correspond, in Figure 5.1, to the piecewise-linear graph with a *solid* line connecting its points. Similarly, two other graphs are plotted. The graph with *dotted* lines corresponds to use of the standard starting point with components scaled by -10.0. And the graph with *dashed* lines corresponds to use of the standard starting point with components scaled by $+10.0$.

The three performance graphs of Figure 5.1 raise the following interesting question: *Is there a best CG algorithm?* We will discuss this fundamental issue in Chapter 14, where the new CG family provides the basis for illustrating a new population-based paradigm for optimization.

5.5.4 Conclusions

The subject of conjugate gradients provides a natural bridge between the field of computational linear algebra and linear equation-solving, on the one hand, and the field of nonlinear optimization and nonlinear equation-solving, on the other; see the conference theme and research articles in Adams and Nazareth [1996].

In the nonlinear context it is important to take a deliberately broad view of conjugate gradients and to place the subject within the arena of more general *multistep iterations* that embrace, for example, the PARTAN and heavy-ball algorithms. Collectively, they define the family of nonlinear *CG-related* algorithms.

In basic or simple multistep algorithms, β_k is an *exogenous scalar parameter*, and the linear CG standard; see the last paragraph of Section 5.1—is not reproduced. In nonlinear CG algorithms, β_k is a *scalar-valued function* of other variables, for example, $\mathbf{g}_{k+1}$, $\mathbf{g}_k$, $\mathbf{y}_k$, $\mathbf{d}_k$, as in (5.7)–(5.8). The resulting techniques conform to the linear CG standard, and they number among the simplest and most elegant algorithms of computational nonlinear optimization.

From a *theoretical* standpoint, nonlinear CG-related algorithms are fascinating *mathematical objects* that are of intrinsic interest, and thus well worthy of careful mathematical analysis from the standpoints of convergence and complexity. This subject currently lacks a *comprehensive* underlying theory, and many interesting algorithmic issues remain to be explored within it.

From a *practical* standpoint, CG-related algorithms provide an option for solving optimization problems of high dimension when computer storage is at a premium. They can also be gainfully employed when the assumptions underlying Newton or quasi-Newton algorithms are violated; see the discussion on "simple" algorithms at the beginning of Chapter 9. When used for these purposes, nonlinear CG algorithms often perform in a surprisingly effective way. Thus, they will always have an honored place within the repertoire of an optimization software library.

A marked disjunction currently exists between theory and practice in the nonlinear CG area. As noted in Section 5.5.1, PPR-based algorithms are considered to be the most efficient in practice, yet they have the *least* satisfactory global convergence properties (Powell [1984]). The DY variant, on the other hand, is the most satisfactory in theory (Dai and Yuan [1999]), but in our experience, it is the least efficient[3] on standard test problems. The FR variant has characteristics in between, while HS is akin to PPR.

---

[3]When $\mathbf{y}_k$ is small, the DY update biases $\mathbf{d}_{k+1}$ towards the previous (unsatisfactory) search direction $\mathbf{d}_k$, in contrast to PPR, which biases $\mathbf{d}_{k+1}$ towards the negative gradient vector $\mathbf{g}_{k+1}$.

The two-parameter family (5.10), which embraces the above four key variants, may have a useful role to play in resolving the theory-practice dilemma. Let us consider *conceptual* and *implementable* algorithms, in turn.

When line searches are *exact*, search directions are guaranteed to be directions of descent. Furthermore, $\mathbf{g}_k^T\mathbf{g}_k = \mathbf{d}_k^T\mathbf{y}_k$, and thus a conceptual algorithm based on (5.10) simplifies to a single parameter λ_k-family that embraces the FR and PPR variants. (In the conceptual setting, recall that FR is equivalent to DY and PPR to HS.) The parameter λ_k can be biased, if necessary, towards FR in order to avoid the potential cycling associated with PPR, while simultaneously retaining considerable flexibility to bias the other way at any iteration.

When line searches are permitted to be *inexact*, with their exit criteria governed by the standard weak or strong Wolfe conditions, then it follows that $\mathbf{d}_k^T\mathbf{y}_k > 0$. Thus, the denominator in (5.10) remains positive in an implementable algorithm. The directional derivative at $\mathbf{x}_{k+1}$ will satisfy $\mathbf{g}_{k+1}^T\mathbf{d}_{k+1} < 0$ provided that

$$l_k \equiv \left[\lambda_k(\mathbf{g}_{k+1}^T\mathbf{g}_{k+1}) + (1-\lambda_k)(\mathbf{g}_{k+1}^T\mathbf{y}_k)\right]\left[\mathbf{g}_{k+1}^T\mathbf{d}_k\right]$$

is strictly less than

$$r_k \equiv \left[\mathbf{g}_{k+1}^T\mathbf{g}_{k+1}\right]\left[\mu_k(\mathbf{g}_k^T\mathbf{g}_k) + (1-\mu_k)(\mathbf{g}_{k+1}^T\mathbf{d}_k - \mathbf{g}_k^T\mathbf{d}_k)\right] > 0.$$

Assuming that $\mathbf{d}_k$ is already a direction of descent, it is easily verified that the parameters λ_k and μ_k can be chosen to ensure that $\mathbf{d}_{k+1}$ is a descent direction; for example, as follows:

- $\mathbf{g}_{k+1}^T\mathbf{d}_k < 0$: If λ_k is chosen so that the first square-bracketed term of l_k is positive, and μ_k is chosen anywhere in $[0,1]$, then $l_k < r_k$.
- $\mathbf{g}_{k+1}^T\mathbf{d}_k = 0$: $l_k < r_k$ for any choice of λ_k and μ_k in $[0,1]$.
- $\mathbf{g}_{k+1}^T\mathbf{d}_k > 0$: The first square-bracketed term of l_k approaches the first square bracketed term of r_k as $\lambda_k \to 1$. Because $\mathbf{g}_k^T\mathbf{d}_k < 0$, the second square-bracketed term of l_k is less than the corresponding term of r_k as $\mu_k \to 0$. Thus $l_k < r_k$ as $\lambda_k \to 1$ and $\mu_k \to 0$.

Simultaneously, it is also evident that considerable freedom of choice for the two parameters λ_k and μ_k remains at any iteration, governed by the signs and relative magnitudes of quantities that arise in the expressions for l_k and r_k, for example, $\mathbf{g}_{k+1}^T\mathbf{g}_{k+1}$ and $\mathbf{g}_{k+1}^T\mathbf{y}_k$. This flexibility can be exploited within a variety of possible strategies. Thus, a detailed exploration of the two-parameter family (5.10) would be a worthwhile undertaking, because it offers the opportunity, and perhaps even the promise, of closing the theory-practice gap in the nonlinear CG area.

5.6 Notes

Sections 5.1–5.2: This material in drawn from Nazareth [2001b] and Nazareth [1996c].

Section 5.5: Some of this material is derived from Nazareth [2001c].

6

Gilding the Nelder–Mead Lily

Exploiting "continuities" between a univariate algorithm and a multivariate counterpart can lead to useful insights. In our concluding chapter of Part II, we explore this topic within the specific context of derivative-free, direct search methods for unconstrained minimization; i.e., methods that iteratively update an initial guess of a solution using a few sampled function values along one or more linearly independent directions.

In the univariate case, a popular direct search method is golden-section (GS) search. This algorithm contains no heuristic parameters, is easily shown to be convergent at a linear rate when f is strictly unimodal, and has a close relationship with Fibonacci search, which is known to be optimal in a minimax sense; see Luenberger [1984].

In the multivariate case, the situation is quite different. The most widely used direct search method is the algorithm of Nelder and Mead [1965], which extends the "simplex" method[1] of Spendley et al. [1962]. The inherent simplicity and ease of implementation of the Nelder–Mead (NM) simplex algorithm and the fact that it can cope with *noise* in the function have contributed to its popularity.[2] However, the NM algorithm is heuristic, contains several arbitrary parameters defining its key steps, and

[1]The method is sometimes called the "polytope" method to distinguish it from Dantzig's simplex method for linear programming, with which it has no connection.

[2]In a recent survey of unconstrained minimization software usage, the Nelder–Mead simplex algorithm was found to be one of the most frequently called routines in a well-known optimization library.

possesses little in the way of convergence theory. For a good overview of the NM simplex algorithm, see Bertsekas [1999] and Kelley [1999].

The gap between the undimensional and multidimensional situations can be exploited. In particular, a natural correspondence can be established between a set of lattice points potentially generated by golden-section search and a set of points utilized within an iteration of the NM algorithm. This observation, in turn, leads to a conceptually appealing variant of the latter algorithm that diminishes its heuristic character and provides a rationale for choosing otherwise arbitrary parameters.

The resulting Nelder–Mead variant (NM-GS), which was developed by Nazareth and Tseng [2002], is more amenable to theoretical analysis, and it is possibly also more effective in practice. In particular, convergence of the NM variant can be guaranteed on strictly unimodal univariate functions, because it then reduces essentially to a variant on golden-section search. This is in marked contrast to the approach taken in Lagarias et al. [1998], where considerable effort is expended to show convergence of the (original) Nelder–Mead algorithm in the univariate case and only on strictly convex functions. With the proposed NM-GS algorithm, one obtains convergence in the univariate case, almost for free, on the broader class of strictly unimodal functions. Some numerical experience of Nazareth and Tseng [2002] with the NM-GS algorithm is also given at the end of the chapter.

6.1 Golden-Section Search

Golden-section direct search is summarized in Figure 6.1. Given boundary points A and E such that the minimum of the objective function is contained in the interval defined by them, two points B and C are placed so that $AC/AE = BE/AE = \alpha \equiv 1/\rho$, where $\rho = (\sqrt{5}+1)/2 \approx 1.618$ is the golden ratio. These quantities satisfy $\rho^2 = \rho + 1$, $\alpha^2 = 1 - \alpha$, and $\alpha^3 = 2\alpha - 1$, where $\alpha \approx 0.618$.

If the function value at C is no greater than that at B, then one can reduce the interval containing the minimum to that defined by B and E.

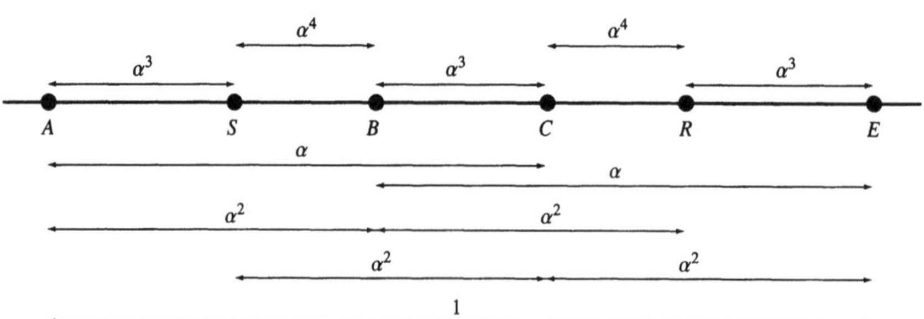

FIGURE 6.1 Golden section search.

The use of the golden ratio ensures that $CE/BE = 1/\rho = \alpha$. Thus, only one additional point, say R, needs to be placed in the new interval with $BR/BE = 1/\rho$. An analogous statement holds for A and C when the value at C is no smaller than that at B, leading to the placement of an additional point S. Then the procedure is repeated. For a detailed description, see Luenberger [1984].

Figure 6.1 depicts the above set of points and relationships between them. Note, in particular, that $AB = BR$; i.e., the point R is an *isometric* reflection of the point A in B along the line AB. Analogously, $CE = SC$. Note also that $AB/BE = \alpha$.

When applied to a strictly unimodal function f, the golden-section procedure converges to the unique minimum of f at a linear rate.

6.2 The Nelder–Mead Algorithm

Figure 6.2 summarizes the Nelder–Mead simplex algorithm for the case $n = 2$. Assume that the numbering of the vertices of the simplex is such that $f_3 \geq f_2 \geq f_1$. The point $\mathbf{x}_r$ is an isometric reflection of the point $\mathbf{x}_3$ in the centroid $\bar{\mathbf{x}}$ of the other two vertices of the simplex, namely, $\mathbf{x}_1$ and $\mathbf{x}_2$. The points $\mathbf{x}_e$, $\mathbf{x}_{oc}$, and $\mathbf{x}_{ic}$ are the usual extrapolation (expansion), outer contraction, and inner contraction points used in the NM algorithm. For a good description, see Bertsekas [1999, pp. 144–148], which we will adapt to present needs in this chapter.

If f is strictly convex and coercive,[3] Lagarias et al. [1998] have shown, via a laborious proof, that the simplices generated by the Nelder–Mead

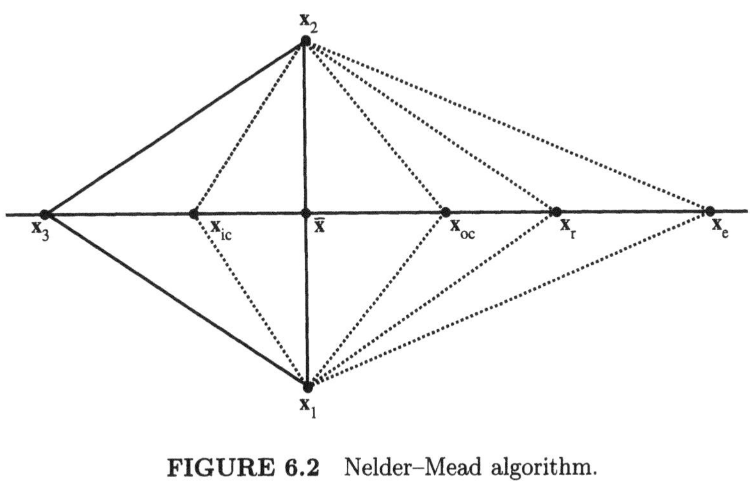

FIGURE 6.2 Nelder–Mead algorithm.

[3]Note that the function is not necessarily differentiable.

algorithm converge to the unique minimizer of f in the unidimensional case of $n = 1$ and that the diameter of the simplices converges to zero in the case of $n = 2$ variables. McKinnon [1998] has constructed a family of strictly convex coercive functions f of $n = 2$ variables on which the simplices generated by the Nelder–Mead algorithm, with particular choices for the initial simplex, converge to a *nonstationary* point.

6.3 Gilding the Lily

Let us superimpose Figure 6.1 on Figure 6.2 and make the identification

$$\mathbf{x}_3, \mathbf{x}_{\rm ic}, \overline{\mathbf{x}}, \mathbf{x}_{\rm oc}, \mathbf{x}_{\rm r}, \mathbf{x}_{\rm e} \quad \text{with} \quad A, S, B, C, R, E,$$

respectively. Thus, we use the choice of parameters within the Nelder–Mead algorithm that is suggested by golden-section search, yielding the NM-GS algorithm below.

At each iteration $k = 0, 1, \ldots$, Algorithm NM-GS updates an n-dimensional simplex, with vertex set $S^{(k)}$, by either moving the worst vertex (highest f-value) in the direction of a centroid of the remaining vertices, with a step size of $1/\rho$ or ρ or 2 or ρ^2, to achieve descent, where ρ is the golden-section ratio of Section 6.1, or by contracting the simplex toward its best vertex.

For any set S of $n+1$ vectors $\mathbf{x}_1, \ldots, \mathbf{x}_{n+1}$ in R^n, define the *diameter* of S to be the quantity

$$d(S) = \max_{i,j} \|\mathbf{x}_i - \mathbf{x}_j\|$$

and let

$$\nu(S) = |\det[\mathbf{x}_2 - \mathbf{x}_1 \quad \cdots \quad \mathbf{x}_{n+1} - \mathbf{x}_1]|/d(S)^n;$$

i.e., $\nu(S)/n!$ is the *volume* of the n-dimensional simplex with vertices given by $(\mathbf{x}_i - \mathbf{x}_1)/d(S)$, $i = 1, \ldots, n+1$, and a diameter of 1. Thus, $\nu(S) = 0$ if and only if the corresponding simplex has one of its interior angles equal to zero or, equivalently, the edges emanating from each vertex of this simplex are linearly dependent.

Let $S\backslash\{\mathbf{x}_{n+1}\}$ denote the remaining set of n vertices of the simplex when $\mathbf{x}_{n+1}$ is omitted.

The following is a restatement of the algorithm of Nazareth and Tseng [2002] that is directly based, for convenience of exposition, on the Nelder–Mead simplex algorithm blueprint given in Bertsekas [1999, pp. 144–148]:

Algorithm NM-GS: Choose any set $S^{(0)}$ of $n+1$ vectors in R^n satisfying $\nu(S^{(0)}) > 0$. For $k = 0, 1, \ldots$, we generate $S^{(k+1)}$ and $\mathbf{x}^{(k)}$ from $S^{(k)}$ as follows:

Step 0. (Initiate) Express $S = S^{(k)} = \{\mathbf{x}_i\}_{i=1}^{n+1}$ in ascending order of function value; i.e., $f_1 = f(\mathbf{x}_1) \leq \cdots \leq f_{n+1} = f(\mathbf{x}_{n+1})$.

$$\overline{\mathbf{x}} = \frac{1}{n}\sum_{i=1}^{n}\mathbf{x}_i, \qquad \overline{f} = \frac{1}{n}\sum_{i=1}^{n} f_i, \tag{6.1}$$

Let $\mathbf{x}^{(k)} = \mathbf{x}_1$.

Step 1. (Isometric Reflection Step) Compute

$$\mathbf{x}_{\mathrm{r}} = \mathbf{x}_{n+1} + 2(\overline{\mathbf{x}} - \mathbf{x}_{n+1}). \tag{6.2}$$

Then compute a new point according to the following three cases:

(1) ($\mathbf{x}_{\mathrm{r}}$ **has min cost**): If $f(\mathbf{x}_{\mathrm{r}}) < f_1$, go to Step 2.

(2) ($\mathbf{x}_{\mathrm{r}}$ **has intermediate cost**): If $f_n > f(\mathbf{x}_{\mathrm{r}}) \geq f_1$, go to Step 3.

(3) ($\mathbf{x}_{\mathrm{r}}$ **has max cost**): If $f(\mathbf{x}_{\mathrm{r}}) \geq f_n$, go to Step 4.

Step 2. (Attempt Expansion) Compute

$$\mathbf{x}_{\mathrm{e}} = \mathbf{x}_{n+1} + \rho^2(\overline{\mathbf{x}} - \mathbf{x}_{n+1}). \tag{6.3}$$

If $f(\mathbf{x}_{\mathrm{e}}) \leq f(\mathbf{x}_{\mathrm{r}})$, then let $S^{(k+1)} = (S\backslash\{\mathbf{x}_{n+1}\}) \cup \{\mathbf{x}_{\mathrm{e}}\}$; else let $S^{(k+1)} = (S\backslash\{\mathbf{x}_{n+1}\}) \cup \{\mathbf{x}_{\mathrm{r}}\}$. Exit.

Step 3. (Use Reflection) Let $S^{(k+1)} = (S\backslash\{\mathbf{x}_{n+1}\}) \cup \{\mathbf{x}_{\mathrm{r}}\}$. Exit.

Step 4. (Perform Contraction) If $f(\mathbf{x}_{\mathrm{r}}) < f_{n+1}$, then compute

$$\mathbf{x}_{\mathrm{oc}} = \mathbf{x}_{n+1} + \rho(\overline{\mathbf{x}} - \mathbf{x}_{n+1}), \tag{6.4}$$

and if $f(\mathbf{x}_{\mathrm{oc}}) < f_{n+1}$, then let $S^{(k+1)} = (S\backslash\{\mathbf{x}_{n+1}\}) \cup \{\mathbf{x}_{\mathrm{oc}}\}$ and Exit;

Otherwise, compute

$$\mathbf{x}_{\mathrm{ic}} = \mathbf{x}_{n+1} + \rho^{-1}(\overline{\mathbf{x}} - \mathbf{x}_{n+1}), \tag{6.5}$$

and if $f(\mathbf{x}_{\mathrm{ic}}) < f_{n+1}$, then let $S^{(k+1)} = (S\backslash\{\mathbf{x}_{n+1}\}) \cup \{\mathbf{x}_{\mathrm{ic}}\}$ and Exit;

Step 5. (Shrink Simplex Toward Best Vertex) Let $S^{(k+1)} = \mathbf{x}_1 + (S - \mathbf{x}_1)/\rho^2$; i.e., the vertex set defined by $\mathbf{x}_1$ and the n points $\mathbf{x}_1 + (\mathbf{x}_i - \mathbf{x}_1)/\rho^2$, $i = 2, \ldots, n+1$, and Exit.

Exit, in the above algorithm, implies a return to Step 0 to begin a fresh cycle. A termination criterion is given in the next section.

For the unidimensional case, the following result concerning the convergence of the (scalar) iterates $x^{(k)}$ is proved in Nazareth and Tseng [2002]:

Proposition 6.1: Assume that $n = 1$ and f is strictly unimodal on R. In Algorithm NM-GS, let $\{(S^{(k)}, x^{(k)})\}_{k=0,1,\ldots}$ be the sequence of simplices and iterates generated. Then, either $\{d(S^{(k)})\} \to \infty$ and f has no minimizer, or $\{d(S^{(k)})\} \to 0$ linearly and $\{x^{(k)}\}$ converges to the unique minimizer of f.

For the multivariate case, convergence requires additional safeguarding of the algorithm along lines developed for NM simplex algorithms in Tseng [2000]. For further details, see Nazareth and Tseng [2002].

6.4 Numerical Experiments

In this section we summarize some numerical experience with Algorithm NM-GS obtained by Nazareth and Tseng [2002].

The algorithm was implemented in Matlab and exercised on eight test functions, and the results are tabulated in Table 6.1. The second and fourth functions are from Powell [1964] and Zangwill [1967], respectively, and the other test functions are nonlinear least-squares problems described in Moré et al. [1981], namely, their problems numbered 1, 11–14, 16.

For each test function, the initial simplex was constructed by taking the starting vector used in the above references and adding to this vector the ith unit coordinate vectors in R^n for $i = 1, \ldots, n$. The algorithm was terminated when $d(S)$ and the ∞-norm of the vector, whose ith component is given by

$$(f_{i+1} - f_1)/\|\mathbf{x}_{i+1} - \mathbf{x}_1\|, \tag{6.6}$$

were both below 10^{-3}, where $S = \{\mathbf{x}_1, \ldots, \mathbf{x}_{n+1}\}$ denotes the current simplex. This ensures that $\nabla f(\mathbf{x}_1) \approx 0$ upon termination, and as can be seen from Table 6.1, the corresponding f-value is with high accuracy close to the global minimum. For comparison, the Nelder–Mead algorithm as interpreted from the original paper of Nelder and Mead [1965] was implemented

TABLE 6.1. Performance of Algorithms NM and NM-GS.

	Algorithm NM		Algorithm NM-GS	
Function	#f-eval.	f-value	#f-eval.	f-value
Powell1 ($n = 4$)	329	$1.5 \cdot 10^{-6}$	183	$4.1 \cdot 10^{-9}$
Powell2 ($n = 3$)	95	−3.0000	90	−3.0000
Rosenbrock ($n = 2$)	159	$1.9 \cdot 10^{-7}$	182	$1.6 \cdot 10^{-8}$
Zangwill ($n = 3$)	86	$3.1 \cdot 10^{-7}$	95	$2.1 \cdot 10^{-7}$
Gulf ($n = 3$)	382	$7.5 \cdot 10^{-12}$	440	$1.1 \cdot 10^{-12}$
Box ($n = 3$)	193	$2.8 \cdot 10^{-10}$	255	$1.6 \cdot 10^{-10}$
Wood ($n = 4$)	807	$2.6 \cdot 10^{-7}$	601	$6.7 \cdot 10^{-8}$
Brown-Dennis ($n = 4$)	298	85823	322	85822
Total	2349	—	2168	—

in Matlab using the same initial simplex and termination criteria as above, and run on the same test functions.

For each algorithm in the table, the first column gives the number of times that f was evaluated upon termination, and the second column gives the value of f at the best vertex upon termination. Algorithm NM-GS performs substantially better than the Nelder–Mead algorithm on two functions (Powell1 and Wood), substantially worse on two functions (Gulf and Box), and about the same on the others. The total number of function calls for the NM-GS algorithm was approximately 10 percent better than the total for the NM algorithm. Thus, while the NM-GS algorithm has nicer theoretical properties than the NM algorithm, their performance seems to be comparable, with the former on balance being slightly superior to the latter on the set of test problems used in these experiments.

6.5 Summary

The theme of this chapter has been the interplay between the golden-section direct search algorithm for univariate optimization and the Nelder–Mead direct search algorithm. The resulting NM-GS algorithm described above is promising in theory and practice. It is obviously a variant of the NM simplex algorithm, because it makes particular choices for expansion, inner contraction and outer contraction parameters. When NM-GS is restricted to the univariate case, it is a variant of golden-section search, because the points that it generates lie on the lattice of points that could potentially be generated by a golden-section procedure with a suitably chosen starting interval. The NM-GS algorithm converges at a linear rate when $n = 1$.

Another example of the overall theme of this chapter can be found in the recent UObyQA algorithm of Powell [2000] for unconstrained optimization by quadratic approximation, achieved by updating Lagrange interpolation functions. The "continuity" between this algorithm and the well-known quadratic-fit minimization algorithm in one dimension is evident; see expression (4.4). Thus, when implemented appropriately, algorithm UObyQA will exhibit a superlinear rate of convergence in the univariate case, with order approximately 1.3.

6.6 Notes

Sections 6.1–6.4: The material for this chapter is based on Nazareth and Tseng [2002].

Part III

Choosing the Right Diagonal Scale

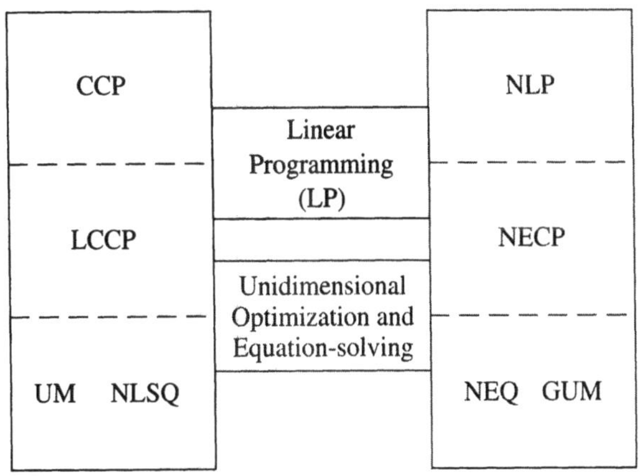

7
Historical Parallels

By means of simple transformations, in particular, the introduction of new variables and nonnegative bounds, an arbitrary linear program can always be restated in *standard form*, namely,

$$\begin{aligned} &\text{minimize } \mathbf{c}^T\mathbf{x} \\ &\text{s.t. } \mathbf{A}\mathbf{x} = \mathbf{b}, \quad \mathbf{x} \geq \mathbf{0}, \end{aligned} \tag{7.1}$$

where $\mathbf{A}$ is an $m \times n$ matrix with $m \leq n$, $\mathbf{x}$ is a vector of n unknowns, and the other quantities are vectors of matching dimensions. The feasible region, formed by the intersection of the m hyperplanes defined by the rows of $\mathbf{A}$ and the nonnegative orthant, is a bounded or unbounded convex polyhedral set in n dimensions.

Consider the following very simple instance in three variables: Minimize a linear objective function $\mathbf{c}^T\mathbf{x}$, $\mathbf{x} \in R^3$, where the solution must satisfy a single linear equality constraint $\mathbf{a}^T\mathbf{x} = b$ in nonnegative variables $\mathbf{x} \geq \mathbf{0}$. The vectors $\mathbf{x} \in R^3$ and $\mathbf{a} \in R^3$ and the scalar b are given, and additionally, the elements of $\mathbf{a}$ and b are restricted to *positive* numbers. The solution of this linear program lies in the intersection of the plane defined by the equality constraint and the nonnegative orthant as depicted in Figure 7.1. The objective function decreases along the direction $(\mathbf{x}^{(1)} \rightarrow \mathbf{x}^{(2)})$, and the solution of the linear program is at the vertex $\mathbf{x}^* \in R^3$.

We will use this simple example to illustrate the linear programming algorithms introduced in this chapter, and a similar example in Chapter 10.

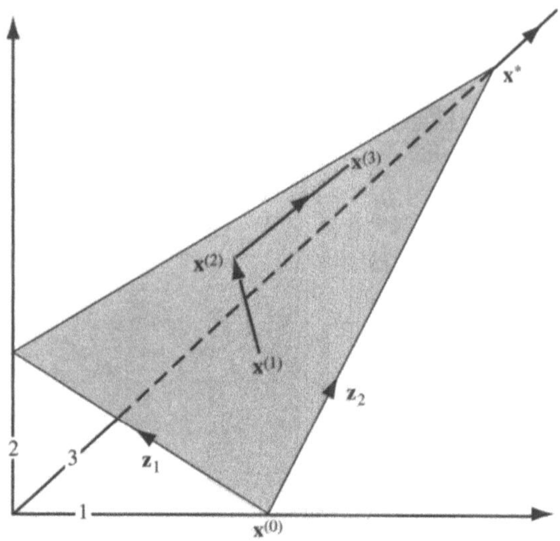

FIGURE 7.1 LP example.

7.1 The Simplex Algorithm

The simplex algorithm, when applied to the LP example of Figure 7.1, proceeds as follows: Start at any vertex of the triangular feasible region, say $\mathbf{x}^{(0)}$, find an edge along which the objective function decreases, and move along it to an adjacent vertex. Then repeat until no further progress can be made, which can happen only at the optimal solution. If the three elements of $\mathbf{a}^T$ are denoted by $[\beta, \eta_1, \eta_2]$, then the vertex $\mathbf{x}^{(0)}$ and the vectors $\mathbf{z}_1$ and $\mathbf{z}_2$ along the edges to the two adjacent vertices are as follows:

$$\mathbf{x}^{(0)} = \begin{bmatrix} \beta^{-1}b \\ 0 \\ 0 \end{bmatrix}, \quad \mathbf{z}_1 = \begin{bmatrix} -\beta^{-1}\eta_1 \\ 1 \\ 0 \end{bmatrix}, \quad \mathbf{z}_2 = \begin{bmatrix} -\beta^{-1}\eta_2 \\ 0 \\ 1 \end{bmatrix}, \tag{7.2}$$

where $\beta^{-1}b > 0$.

The scalar products between the vector $\mathbf{c}$ and the vectors $\mathbf{z}_1$ and $\mathbf{z}_2$ define the directional derivatives along the edges, and one can proceed along any edge whose directional derivative is negative. For example, one could choose the *steepest edge* as follows: Normalize each vector $\mathbf{z}_i$, $i = 1, 2$, by dividing it by its length $||\mathbf{z}_i||$, compute the directional derivatives along the normalized vectors, and choose the most negative. In Figure 7.1, this would be, say, the edge $\mathbf{z}_1$. Note also that the vectors $\mathbf{z}_1$ and $\mathbf{z}_2$ are orthogonal to $\mathbf{a}$; i.e., the matrix $\mathbf{Z} = [\mathbf{z}_1, \mathbf{z}_2]$ spans the null space of the matrix $\mathbf{A} \equiv \mathbf{a}^T = [\beta, \eta_1, \eta_2]$, and hence $\mathbf{AZ} = \mathbf{0}$. Proceed along an improving edge to a new vertex. In order to obtain an identical situation at the new vertex to the one just described, simply reorder the variables. Then repeat the procedure.

The description of the simplex algorithm for more general m and n in the linear program (7.1) is no different, in essence, from the description on the earlier simple example.[1] Partition the columns of the LP matrix $\mathbf{A}$ into $\mathbf{A} = [\mathbf{B}, \mathbf{N}]$, where $\mathbf{B}$ is an $m \times m$ *basis* matrix assumed to be nonsingular and defined by the first m columns of $\mathbf{A}$, and $\mathbf{N}$ is an $m \times (n-m)$ matrix consisting of the remaining columns of $\mathbf{A}$. (In the earlier example, $m = 1$, $n = 3$, the matrix $\mathbf{B}$ corresponds to the 1×1 matrix β, and $\mathbf{N}$ to the 1×2 matrix $[\eta_1, \eta_2]$.) The starting vertex is given by a vector $\mathbf{x}^{(0)}$, and the edges leading from it are defined by the $n - m$ columns of a matrix $\mathbf{Z}$ as follows:

$$\mathbf{x}^{(0)} = \begin{bmatrix} \mathbf{B}^{-1}\mathbf{b} \\ \mathbf{0} \end{bmatrix}, \qquad \mathbf{Z} = \begin{bmatrix} -\mathbf{B}^{-1}\mathbf{N} \\ \mathbf{I} \end{bmatrix}, \tag{7.3}$$

where $\mathbf{I}$ denotes the $(n-m) \times (n-m)$ identity matrix and $\mathbf{0}$ denotes a zero vector of appropriate dimension. We assume that the basis matrix $\mathbf{B}$ is such that $\mathbf{B}^{-1}\mathbf{b} \geq \mathbf{0}$. The *directional derivatives* along edges are given by the elements of the vector $\mathbf{c}^T\mathbf{Z}$. Again one can proceed along any edge with a negative directional derivative to an improving vertex. Note that $\mathbf{Z}$ spans the null space of the matrix $\mathbf{A}$ and that the operations of the simplex method now involve inversion of the basis matrix, which is usually handled implicitly by forming and updating matrix factorizations and solving associated systems of linear equations.

At a vertex, the set of m equality constraints in (7.1) and the set of $(n-m)$ nonnegative bound constraints that hold as equalities (the corresponding nonbasic variables are fixed at value zero) then jointly comprise a set of n constraints that are said to be *active* at the point $\mathbf{x}^{(0)}$. The corresponding directional derivatives along edges $\mathbf{c}^T\mathbf{Z}$ are the *Lagrange multipliers* that correspond to the $(n-m)$ active bound constraints at $\mathbf{x}^{(0)}$. In the simplex algorithm, one of these active constraints, whose Lagrange multiplier (directional derivative, *reduced cost*) has a negative sign, is relaxed, and a move is then made along the corresponding edge to a new vertex, again defined by a set of n active constraints. In this interpretation, the simplex algorithm is seen to be a *specialized active-set algorithm* for

---

[1]Now Figure 7.1 is no longer a "pictorial representation" of a particular linear program of the form (7.1), but rather a "compound symbol" or *icon* that *denotes* this linear program, in general. Thus, the triangular feasible region denotes a bounded or unbounded polyhedral feasible region in R^n, which lies within an affine space of dimension $n - m$, namely, the intersection of the m hyperplanes corresponding to the equality constraints of (7.1). Each "axis" in Figure 7.1 denotes a subspace, say S, defined by m variables that remain when a subset of $(n-m)$ nonbasic variables are set to value zero. These m basic variables must also satisfy the m equality constraints, and their values are therefore determined. When their values are nonnegative, then the basic and nonbasic variables define a vertex of the feasible region that lies in the subspace S.

solving the LCCP problem (1.5) of Chapter 1; see also Figure 1.2. Thus the simplex algorithm can be approached in a very natural way from the left-hand, or convexity, side of the watershed portrayed in Figure 1.2. For a detailed discussion; see Nazareth [1987, Chapter 2].

Under nondegeneracy assumptions, the simplex algorithm is easily shown to converge to the optimal solution of a linear program in a finite number of iterations. On a degenerate linear program, the very elegant rule proposed by Bland [1977]—assume a fixed ordering of the variables and resolve ties in the entering or exiting variables by always choosing the candidate with the *smallest* index in the ordering—is the simplest among several techniques that are available for selecting entering and exiting variables to guarantee convergence of the algorithm. For details, see, for example, Chvatal [1983].

7.2 The Affine-Scaling Algorithm

Now consider an alternative approach. Suppose that the current approximation to the solution is at an interior point of the feasible region, say $\mathbf{x}^{(1)} > \mathbf{0}$ in Figure 7.1. If $\mathbf{P}$ denotes the orthogonal projector into the plane defined by the equality constraint, then $\mathbf{d} = \mathbf{P}(-\mathbf{c})$, the projected negative-gradient direction, would be a natural choice for the improving direction. This direction is also depicted in Figure 7.1. The projection matrix is

$$\mathbf{P} = \mathbf{I} - \mathbf{A}^T(\mathbf{A}\mathbf{A}^T)^{-1}\mathbf{A} = \mathbf{Z}(\mathbf{Z}^T\mathbf{Z})^{-1}\mathbf{Z}^T, \tag{7.4}$$

where $\mathbf{A}$ is defined in (7.1) and $\mathbf{Z}$ is a matrix of full rank spanning the null space of $\mathbf{A}$, for example, the matrix defined in expression (7.3). One can then compute the step length to the boundary, which is given by the largest value of α for which all components of $\mathbf{x} + \alpha\mathbf{d}$ remain nonnegative. Choose a step length slightly smaller in order to remain in the interior of the feasible polytope; i.e., to ensure that all components remain strictly positive. Suppose this gives the new iterate $\mathbf{x}^{(2)}$. For the particular choice of $\mathbf{x}^{(1)}$ shown in Figure 7.1, good progress will be made during the first iteration, but there is no point in repeating the procedure at $\mathbf{x}^{(2)}$. The same projected negative-gradient direction as before would be obtained, and the iterates would jam near the boundary. The affine scaling method is based on a novel yet incredibly simple modification of this approach—terrific ideas are often very simple with hindsight! This is as follows: *Change the metric or measure of distance by a rescaling of variables.* Suppose $\mathbf{x}^{(k)}$, $k \geq 2$, denotes the current iterate and let $\mathbf{X}_k$ denote the *diagonal* matrix whose diagonal elements are the components of $\mathbf{x}^{(k)}$; i.e., $\mathbf{X}_k = \text{diag}\left[x_1^{(k)}, x_2^{(k)}, x_3^{(k)}\right]$. Make the linear transformation or scaling of variables $\mathbf{x} = \mathbf{X}_k\tilde{\mathbf{x}}$. Thus the transformed linear program is as follows: Minimize $\tilde{\mathbf{c}}^T\tilde{\mathbf{x}}$ subject to $\tilde{\mathbf{A}}\tilde{\mathbf{x}} = \mathbf{b}$, $\tilde{\mathbf{x}} \geq \mathbf{0}$, where $\tilde{\mathbf{c}} = \mathbf{X}_k\mathbf{c}$ and $\tilde{\mathbf{A}} = \mathbf{A}\mathbf{X}_k$. Compute the projected direction $\tilde{\mathbf{d}}$ as we did above, but now in the transformed space; i.e., $\tilde{\mathbf{d}} = \tilde{\mathbf{P}}(-\tilde{\mathbf{c}})$. In the original space this corresponds to the direction $\mathbf{d} = \mathbf{X}_k\tilde{\mathbf{d}} = -\mathbf{X}_k\tilde{\mathbf{P}}\mathbf{X}_k\mathbf{c}$, and in terms of the quantities that define the original linear program we have

$$\mathbf{d} = -\mathbf{X}_k\left[\mathbf{I} - \mathbf{X}_k\mathbf{A}^T(\mathbf{A}\mathbf{X}_k^2\mathbf{A}^T)^{-1}\mathbf{A}\mathbf{X}_k\right]\mathbf{X}_k\mathbf{c}. \tag{7.5}$$

At $\mathbf{x}^{(2)}$, the net effect is to bend the search direction away from the boundary of the feasible region as shown, so that greater progress is possible. Take an improving step subject to remaining in the interior of the feasible region. Then repeat the procedure. This, in essence, defines the primal affine-scaling algorithm, a particular realization of the underlying affine-scaling *method.* Again, under appropriate nondegeneracy assumptions (and even in their absence, but then proofs are harder to obtain), the sequence of iterates can be shown to converge; i.e., $\mathbf{x}^{(k)} \to \mathbf{x}^*$.

The statement of the affine scaling algorithm, in general, is no more complicated than the one given for the earlier example. The search directions are defined by (7.5), where $\mathbf{X}_k$ is an $n \times n$ diagonal matrix with diagonal elements defined by the components of $\mathbf{x}^{(k)} > \mathbf{0}$. But now the quantity $(\mathbf{A}\mathbf{X}_k^2\mathbf{A}^T)^{-1}$ is an $m \times m$ matrix, not just a scalar. Implicitly or explicitly, the operations of the affine-scaling algorithm involve matrix inversion, as in the simplex algorithm. The remainder of the mathematical algorithm for choosing the next iterate is as described above for the example of Figure 7.1.

An alternative expression for $\mathbf{d}$ can be given in terms of $\mathbf{Z}$. Let $\tilde{\mathbf{Z}}$ be a matrix of full rank that spans the null space of $\tilde{\mathbf{A}}$. Clearly, $\mathbf{Z}$ and $\tilde{\mathbf{Z}}$ are related by $\tilde{\mathbf{Z}} = \mathbf{X}_k^{-1}\mathbf{Z}$. Then, using (7.4) in the transformed space and reverting to the original variables gives

$$\mathbf{d} = -\mathbf{Z}(\mathbf{Z}^T\mathbf{X}_k^{-2}\mathbf{Z})^{-1}\mathbf{Z}^T\mathbf{c}. \tag{7.6}$$

This alternative form for $\mathbf{d}$ has advantages for computation, as we will see in the illustration of the next section.

As with the simplex algorithm, observe that the affine-scaling approach is rooted in the left-hand, convexity, side of the watershed of Figure 1.2. Specifically, it draws on the Newton–Cauchy framework of Chapter 2 for its foundation. We will continue the development along these lines in the next chapter.

7.3 Numerical Illustration

We now give a numerical illustration of the foregoing mathematical algorithms, simplex and affine-scaling, using the classical test problems of Kuhn and Quandt [1963]; see also Avis and Chvatal [1978] and Chvatal [1983]. These problems are defined as follows:

$$\text{minimize } -\mathbf{e}^T\mathbf{x}_N$$
$$\text{s.t. } \mathbf{x}_B + \mathbf{N}\mathbf{x}_N = 10^4\mathbf{e}, \quad \mathbf{x}_B \geq \mathbf{0}, \mathbf{x}_N \geq \mathbf{0},$$

where $\mathbf{e}$ is a vector of all 1's, and $\mathbf{N}$ is an $m \times m$ matrix with integer elements chosen at random in the range $1, \ldots, 1000$. It is convenient to place the columns corresponding to the slack variables at the front of the LP matrix and to use them to define the initial basis matrix, hence the choice of notation $\mathbf{x}_B$. The values of m are chosen to be 10, 20, 30, 40, 50. The starting point is defined by setting each component of $\mathbf{x}_N^{(0)}$ to $10/m$. The corresponding slack variables $\mathbf{x}_B^{(0)}$ are then feasible, and the starting objective value is $z_{\text{init}} = -10.0$, regardless of the size of m.

We will illustrate the relative performance of the simplex algorithm and the affine-scaling algorithm on the above small, dense problems. The two algorithms were implemented as follows:

Simplex Algorithm: This is the standard tableau form of the primal simplex algorithm; see Dantzig [1963] or Dantzig and Thapa [1997]. For dense LP problems like the Kuhn-Quant problems, where the total number of columns (including slacks) is no more than double the total number of rows, an iteration of the tableau simplex algorithm is *cheaper* than an iteration of the revised simplex algorithm; see Chvatal [1983, p. 114].

The initial basis matrix is taken to be the identity matrix, corresponding to the slack variables. The implementation also incorporates a simple extension that permits the algorithm to start at an *interior* point; i.e., the starting point is not required to be a basic feasible solution. For details; see Nazareth [1989, Section 2.2-1].

Affine-Scaling Algorithm: This implements the null-space form (7.6) of the affine-scaling direction where $\mathbf{Z}$ is defined by (7.3) with $\mathbf{B} = \mathbf{I}$. Define the $m \times m$ positive definite diagonal matrix $\mathbf{X}_B$ to have diagonal elements $\mathbf{x}_B^{(0)}$, and the $m \times m$ positive definite diagonal matrix $\mathbf{X}_N$ to have diagonal elements $\mathbf{x}_N^{(0)}$. Let $\mathbf{X}_k$ in (7.6) be the diagonal matrix with elements given by $\mathbf{X}_B$ and $\mathbf{X}_N$. Then the expression (7.6) defining the search direction can be rewritten as follows:

$$\mathbf{d} = \begin{bmatrix} -\mathbf{N} \\ \mathbf{I} \end{bmatrix} \left[\mathbf{N}^T \mathbf{X}_B^{-2} \mathbf{N} + \mathbf{X}_N^{-2}\right]^{-1} \mathbf{e}. \tag{7.7}$$

Let $\mathbf{d}^T = [\mathbf{d}_B^T, \mathbf{d}_N^T]$. Then $\mathbf{d}_N$ is the solution of the following system of linear equations:

$$\left[\mathbf{N}^T \mathbf{X}_B^{-2} \mathbf{N} + \mathbf{X}_N^{-2}\right] \mathbf{d}_N = \mathbf{e}, \tag{7.8}$$

and

$$\mathbf{d}_B = -\mathbf{N}\mathbf{d}_N.$$

The linear system (7.8) can be reformulated as

$$\left[\mathbf{I} + \mathbf{X}_N \mathbf{N}^T \mathbf{X}_B^{-2} \mathbf{N} \mathbf{X}_N\right] \hat{\mathbf{d}}_N = \hat{\mathbf{e}}, \tag{7.9}$$

where $\mathbf{d}_N = \mathbf{X}_N \hat{\mathbf{d}}_N$ and $\hat{\mathbf{e}} = \mathbf{X}_N \mathbf{e}$. This is equivalent to a *diagonal preconditioning* of the linear system (7.8).

The next iterate is obtained as follows. Compute the maximum step, say $\bar{\alpha}$, that can be taken along the direction $\mathbf{d} = (\mathbf{d}_B, \mathbf{d}_N)$, without overreaching a nonnegative bound constraint and thus violating feasibility. Then the step from the current to the next iterate is taken to be $\alpha\mathbf{d}$, where $\alpha = 0.9\bar{\alpha}$. The vector $\mathbf{x}^{(0)} = \left(\mathbf{x}_B^{(0)}, \mathbf{x}_N^{(0)}\right)$ is updated to be the new iterate, and the procedure is then repeated.

The above linear system of equations (7.8), which is used to obtain $\mathbf{d}_N$, is solved in two different ways:

- *Cholesky Factorization:* The matrix defining the linear system (7.9), say $\mathbf{M}$, is formed explicitly, and the system is then solved by forming the Cholesky factorization of $\mathbf{M}$, followed by forward and back substitution. The linear algebra routines are a direct encoding of the procedures given in Golub and Van Loan [1983, pages 53 and 89].
- *CG:* At any iteration, say k, the system (7.9) is solved by the conjugate gradient algorithm given in Golub and Van Loan [1983, p. 374], with no preconditioning beyond the explicit diagonal preconditioning mentioned above. (See also our Section 5.1 for an overview of the linear CG algorithm.) The algorithm is initiated at the origin, and the matrix $\mathbf{M}$ is *not* formed explicitly; i.e., the matrices defining it are used in matrix–vector operations within the CG algorithm. The CG algorithm is terminated after ncg_k steps, where $ncg_1 = \sqrt{m}$ and subsequently $ncg_{k+1} \leftarrow ncg_k + k$. In addition, the CG algorithm is terminated if the residual of the system of equations, say $\mathbf{r}$, falls below a small quantity, specifically, $\mathbf{r}^T\mathbf{r} \leq 10^{-3}$. The CG algorithm always produces a direction of descent, even when it is terminated prematurely. For further discussion; see Nazareth [1989].

We will refer to the two variants of the affine-scaling algorithm, implemented as outlined above, as Dikin–Cholesky and Dikin–CG, respectively.

The affine scaling algorithm generates an infinite sequence of iterates, and it is sensitive to the stopping criterion. Since the simplex algorithm has finite termination, it is used as a benchmark in the numerical illustration as follows. A choice is made for m from the five choices listed above, and the test problem is generated. The simplex algorithm is run to completion on the test problem, yielding an optimal solution. The objective values of the iterates prior to termination, along with the computer time to reach them, are examined, and at the iterate, say K, where all figures in the objective value up to the second after the decimal point have settled down to their final optimal value, the run time T is recorded. (Subsequent iterates are ignored.) In this way, we avoid occasions, if any, where the simplex algorithm dawdles near the optimum, yet we also avoid premature termination. Let z_{opt} denote the objective value for simplex iterate K.

The Dikin–Cholesky algorithm is then run on the test problem until the completion of the first iterate, say J, for which the total run time exceeds T. The objective value attained is recorded, and if the run time is substantially in excess of T, the objective value obtained is averaged with the objective value at the previous iterate, where the run time is obviously less than T. Let us denote this recorded value by z_f. Then, in the column of Table 7.1 corresponding to this algorithm, we report the *relative error* expressed as a percentage, namely,

$$R = \frac{z_f - z_{\text{opt}}}{z_{\text{init}} - z_{\text{opt}}} 100\%. \tag{7.10}$$

This is the number in boldface in the table. (Obviously, the corresponding quantity R is zero for the simplex algorithm, which is being used as a benchmark in these experiments.) The table also reports the number of iterations as the first number of each entry in the table. If z_f in the Dikin algorithm is obtained by averaging, then the corresponding iteration number is recorded as $(J + (J - 1))/2$. Otherwise, it is the whole number J.

The entire procedure just described is then repeated for the affine scaling algorithm Dikin–CG, which was implemented as outlined above. In this case, the table also reports a third number, the average number of CG steps taken; i.e., the total number of CG steps over all relevant iterations divided by the number of iterations of the affine scaling algorithm. For the smallest test problem, $m = 10$, we set $ncg_1 = 1$ in place of the initialization stated above.

The algorithms were implemented in Microsoft Basic using double-precision arithmetic. For a given value of m, each algorithm is initiated from precisely the same starting point and run on precisely the same test problem. Since the implementation of each algorithm was done in a uniform way, essentially a direct encoding of the underlying formulae defining the algorithms, and the linear algebra was also done without embellishment of the Golub–Van Loan procedures quoted above, the experiment is a very uniform one, *without bias* toward a particular algorithm.

The Dikin–Cholesky algorithm performs surprisingly well in these experiments, given the small number of iterations permitted to it, and the Dikin–CG algorithm with diagonal preconditioning and inexact computa-

TABLE 7.1. Performance of simplex and affine-scaling algorithms.

Dimension	Simplex	Dikin–Cholesky	Dikin–CG
10	12, **0.**	2.5, **17.8**	3.5, **11.52**, 3.3
20	30, **0.**	3.0, **7.40**	5.0, **1.44**, 7.6
30	37, **0.**	2.0, **28.0**	5.0, **2.60**, 8.0
40	85, **0.**	3.5, **6.46**	7.5, **0.67**, 12.6
50	128, **0.**	4.0, **3.20**	10.0, **0.48**, 14.6

tion of search directions is seen to be a promising competitor. We emphasize that the foregoing are simply level-1 experiments with mathematical algorithms and contrived examples—for terminology see Section 15.5.1—and one cannot extrapolate performance to other, more realistic, test problems. Nevertheless, the experiments represent a useful illustration of the behavior of the simplex and affine-scaling algorithms on a set of small and dense classical test problems.

7.4 Historical Parallels

It is when one turns to *practical* linear programs and formulates the above methods into effective *numerical* algorithms that linear programming comes into its own. In a practical linear program (7.1), the matrix **A** could have thousands of rows and tens of thousands of columns and would generally be very sparse; i.e., a typical column would contain only a few nonzero elements, often between 5 and 10. (In a practical linear program, the nonnegativity bounds are also often replaced by upper and lower bounds on the variables.) Two considerations are then of paramount importance. How many iterations, as a function of problem dimensions, do our methods take on typical problems? How efficient can each iteration be made? These *complexity* considerations are what distinguish the *numerical* algorithm from the *mathematical* algorithm of the previous sections.

In the case of the simplex method, the *mathematical* algorithm, described in terms of vertex following, was proposed independently, at least three different times. The contributions of G.B. Dantzig to the *numerical* algorithm are best described in his own words; see Dantzig [1983]:

> It is my opinion that any well trained mathematician viewing the linear programming problem in the row geometry of the variables would have immediately come up with the idea of solving it by a vertex descending algorithm as did Fourier, de la Valee Poussin and Hitchcock before me—each of us proposing it independently of the other. I believe, however, that if anyone had to consider it as a practical method, as I had to, he would quickly have rejected it on intuitive grounds as a very stupid idea without merit. My own contributions towards the discovery of the simplex method were (1) independently proposing the algorithm, (2) initiating the software necessary for its practical use, and (3) observing by viewing the problem in the geometry of the columns rather than the rows that, contrary to geometric intuition, following a path on the outside of the convex polyhedron might be a very efficient procedure.

Item (3) above represents one of the early contributions to *complexity analysis.* In practice, the typical number of iterations of the simplex algorithm increases proportionally to m, with a very small constant of proportionality (for example, between 1.5 and 3). The arguments that led Dantzig to this fact were later formalized by him; see, for example, Dantzig [1980]. The "LP technology" initiated by Dantzig relies on and, indeed, stimulated the development of appropriate techniques for solving large and very sparse systems of linear equations. These techniques are based on factorization of the matrix, and they implicitly perform the basis matrix inversion involved in (7.3). (They are called direct methods because they operate directly on the basis matrix $\mathbf{B}$ in order to obtain a representation of it that permits very efficient solution of the associated systems of linear equations.) These then were the two key contributions of Dantzig to the simplex method, in addition to proposing the mathematical algorithm independently of earlier historical antecedents, as mentioned in the above quotation. *It is important to appreciate the quantum leap from a viable mathematical algorithm to a viable numerical algorithm*, which is why Dantzig's name is synonymous with the simplex method. Subsequent to the acceptance of the simplex method as the workhorse of linear programming, there was an explosion of development that led to numerous variants, for example, the dual simplex algorithm and the self-dual simplex homotopy algorithm. It is also worth noting that linear programming as we know it today would not have been possible without the digital computer, and developments in the two fields advanced together in tandem; see Dantzig [1985].

The affine scaling method has its roots in the steepest descent method of Cauchy [1847]. The mathematical algorithm was first proposed by the Russian mathematician Dikin [1967], a student of the legendary mathematician Leonid Kantorovich, whose name is also associated with the origins of linear programming; see Kantorovich [1939]. However, Dikin's algorithm was all but forgotten. It reemerged as a viable numerical technique when it was independently discovered by several researchers as an affine-scaling variant of the projective-scaling algorithm of Karmarkar [1984]—the simplification from projective to affine was obviously in the air—thereby widening the repertoire of LP solution techniques that until then had been largely dominated by the simplex method. As was the case with Dantzig's development of the simplex method, Karmarkar's key contributions were to the *numerical* interior-point algorithm, these being twofold. First, he discovered that the number of iterations required to reach the solution was typically very small. It has been described as growing with the logarithm of the number of variables n. Secondly, Karmarkar spearheaded the development of techniques for increasing the efficiency of each iteration. Interestingly enough, the use of iterative techniques was strongly championed by Karmarkar, although direct techniques remain the most widely used for implementing interior-point algorithms in practice. The full potential of iterative techiques for the associated computational linear algebra has still to be fully tapped. And it is worth noting that iterative techniques are more amenable to computation on a *multiprocessor* machine, so that the LP technology of affine-scaling and interior methods, in general, is likely to develop in tandem with the development of increasingly powerful computers of this type. Again, variants of the affine-scaling algorithm were quick to follow, in particular, a numerical version of the dual affine-scaling algorithm; see Alder, Karmarkar, Resende, and Veiga [1989]. This was the affine-scaling variant of the projective-scaling algorithm that was first investigated on realistic problems, and it served to highlight the efficiency of the interior-point approach. Another key variant, discovered soon afterwards, was the primal–dual affine-scaling algorithm, which will be discussed further in the next chapter.

The historical parallels between the development of the simplex method of Dantzig and and the interior-point method of Karmarkar are striking. *But there is an important difference.* After making the initial breakthrough in formulating and recognizing the enormous practical importance of the linear programming model and discovering the numerical simplex algorithm, Dantzig continued to play a key role in developing many other fundamental concepts of optimization: duality and its relationship to optimality conditions, the self-dual simplex homotopy algorithm, the Dantzig–Wolfe decomposition principle and algorithm, stochastic programming with recourse, the use of importance sampling and parallel processors, and so on. This leadership, extending over a period of almost five decades, is the reason that he is universally and affectionately known as the *father of mathematical programming*. The "golden age" of optimization that was ushered in by the breakthroughs of the nineteen forties—the Dantzig Modeling-and-Algorithmic Revolution—*subsumes* the more recent algorithmic revolution initiated by Karmarkar [1984].

For historical background and a fascinating account of the origins of linear programming; see the foreword of Dantzig and Thapa [1997], and for an extensive bibliography of G.B. Dantzig's contributions; see pp. 368–384 of the same reference. For a wide-ranging overview of the entire field of optimization; see the recently published *Encyclopedia of Optimization*, Floudas and Pardalos [2001].

7.5 Notes

Sections 7.1, 7.2, 7.4: An earlier version of this material can be found in Nazareth [1990].

Section 7.3: The numerical experiments and discussion in this section are derived from Nazareth [1989].

8

LP from the Newton–Cauchy Perspective

Consider again the linear programming problem in standard form and its dual, namely,

$$\begin{array}{lll} (P): \text{ minimize } \mathbf{c}^T\mathbf{x} & & (D): \text{ maximize } \mathbf{b}^T\pi \\ \quad \text{s.t. } \mathbf{A}\mathbf{x} = \mathbf{b}, & \text{and} & \quad \text{s.t. } \mathbf{A}^T\pi + \mathbf{v} = \mathbf{c}, \\ \quad \mathbf{x} \geq \mathbf{0}, & & \quad \mathbf{v} \geq \mathbf{0}, \end{array} \tag{8.1}$$

where $\mathbf{x}$, $\mathbf{v}$, and $\mathbf{c}$ are n-vectors, $\mathbf{b}$ and π are m-vectors, and $\mathbf{A}$ is an $m \times n$ matrix with $m \leq n$. We assume that $\mathbf{A}$ is of full rank and that the primal and dual feasible sets have nonempty interiors.

In the previous chapter, Dikin's affine-scaling algorithm for solving the linear program (P) was introduced from the metric-based, or Cauchy, perspective of Section 2.3. We now formulate the Dikin algorithm using the other two main Newton–Cauchy (NC) perspectives of Chapter 2; see Sections 2.2 and 2.4. Within the *linear* programming context of the primal problem (P), the three NC approaches are shown to be *equivalent* to one another. We then also formulate affine scaling in the setting of the dual problem (D), and in the setting of the primal and dual problems considered simultaneously. The conditions that define the *central path* of a linear program in primal–dual space—a fundamental object of linear programming—emerge as a very natural outcome of our line of development.

In this chapter our focus is on the derivation of the three main forms of affine scaling—primal, dual, and primal–dual—and the *interrelationships* among their corresponding search directions. These directions can be used to formulate interior-point algorithms along lines described and illustrated numerically in the previous chapter. Related issues of convergence and effective implementation of affine-scaling algorithms are considered briefly in the concluding section.

8.1 Primal Affine Scaling

In the discussion that follows, let $\mathbf{x}^{(k)}$ be a feasible interior point of the primal linear program (P); i.e., $\mathbf{x}^{(k)} > \mathbf{0}$ and $\mathbf{A}\mathbf{x}^{(k)} = \mathbf{b}$, and let $\mathbf{X}_k$ be the following positive definite diagonal matrix:

$$\mathbf{X}_k = \operatorname{diag}\left[x_1^{(k)}, \ldots, x_n^{(k)}\right]. \tag{8.2}$$

8.1.1 Model-Based Perspective

In the primal linear program (8.1), the bound constraints $\mathbf{x} \geq \mathbf{0}$ are replaced by a quadratic trust region,

$$\left(\mathbf{x} - \mathbf{x}^{(k)}\right)^T \mathbf{X}_k^{-2} \left(\mathbf{x} - \mathbf{x}^{(k)}\right) \leq 1. \tag{8.3}$$

The trust region lies in R^n_+ and touches each axis, and it is called the *Dikin ellipsoid*. The associated local approximating model, or direction-finding problem (DfP), is the analogue of expressions (2.2) when $\mathbf{H}_k$ is set to the zero matrix, the space is further restricted to satisfy the linear constraints and the trust region is defined by (8.3), and the DfP is thus defined as follows:

$$\begin{aligned} &\text{minimize } \mathbf{c}^T\left(\mathbf{x} - \mathbf{x}^{(k)}\right) \\ &\text{s.t. } \mathbf{A}\mathbf{x} - \mathbf{b} = \mathbf{A}\left(\mathbf{x} - \mathbf{x}^{(k)}\right) = \mathbf{0}, \\ &\left(\mathbf{x} - \mathbf{x}^{(k)}\right)^T \mathbf{X}_k^{-2} \left(\mathbf{x} - \mathbf{x}^{(k)}\right) \leq 1. \end{aligned} \tag{8.4}$$

A step is taken in the direction of the minimizing point of (8.4). To obtain it, let $\mathbf{Z}$ be a matrix of full rank spanning the null space of $\mathbf{A}$ as in (7.3); i.e., $\mathbf{AZ} = \mathbf{0}$, and make the change of variables $\mathbf{x} = \mathbf{x}^{(k)} + \mathbf{Z}\mathbf{w}$, where $\mathbf{w} \in R^{n-m}$. Then the linear constraints in the DfP are satisfied identically, and the transformed model can be restated as follows:

$$\begin{aligned} &\text{minimize } (\mathbf{Z}^T\mathbf{c})^T\mathbf{w} \\ &\text{s.t. } \|\mathbf{w}\|_{\mathbf{Z}^T\mathbf{X}_k^{-2}\mathbf{Z}} \leq 1. \end{aligned}$$

When we note the appropriate correspondences of symbols between the above transformed DfP and (2.3) and, in particular, that $\mathbf{H}_k$ in (2.3) is the zero matrix, we see that the direction toward the minimizing point is given by

$$\mathbf{d}_w = -\left(\mathbf{Z}^T\mathbf{X}_k^{-2}\mathbf{Z}\right)^{-1}\mathbf{Z}^T\mathbf{c}.$$

Reverting to the space of original variables, this corresponds to the search direction

$$\mathbf{d} = -\mathbf{Z}\left(\mathbf{Z}^T\mathbf{X}_k^{-2}\mathbf{Z}\right)^{-1}\mathbf{Z}^T\mathbf{c} \tag{8.5}$$

at the current iterate $\mathbf{x}^{(k)}$. This is expression (7.6) of the previous chapter, and, as discussed there, it can be written in the alternative form (7.5).

8.1.2 Metric-Based Perspective

The metric-based, or Cauchy, approach[1] of Section 2.3 was used to introduce the affine-scaling algorithm in Chapter 7. The diagonal scaling transformation is determined by the components of the current iterate $\mathbf{x}^{(k)}$; i.e., $\tilde{\mathbf{x}} = \mathbf{X}_k^{-1}\mathbf{x}$ with $\mathbf{X}_k$ given by (8.2). A projected steepest-descent direction is defined at the transformed iterate $\mathbf{e} \equiv (1, 1, \ldots, 1)^T = \mathbf{X}_k^{-1}\mathbf{x}^{(k)}$, and after reverting to the original variables, this direction is given by expressions (7.5) and (7.6). Note also that the boundary hyperplanes are invariant; i.e., $\mathbf{u}_j^T\mathbf{x} = 0,\ j = 1, \ldots, n$, where $\mathbf{u}_j$ denotes the jth unit vector, transforms to $\mathbf{u}_j^T\tilde{\mathbf{x}} = 0,\ j = 1, \ldots, n$, under the affine-scaling diagonal transformation.

8.1.3 NC Perspective

Analogously to Section 2.4, in particular, expression (2.18), make a positive definite *quadratic approximating model* or direction-finding problem (DfP) at $\mathbf{x}^{(k)}$ as follows:

$$\begin{aligned} &\text{minimize } \mathbf{c}^T\left(\mathbf{x} - \mathbf{x}^{(k)}\right) + \tfrac{1}{2}\left(\mathbf{x} - \mathbf{x}^{(k)}\right)^T\mathbf{X}_k^{-2}\left(\mathbf{x} - \mathbf{x}^{(k)}\right) \\ &\text{s.t. } \mathbf{A}\left(\mathbf{x} - \mathbf{x}^{(k)}\right) = \mathbf{0}. \end{aligned} \tag{8.6}$$

We will call (8.6) the *Dikin model.* It is obtained by replacing the bounds $\mathbf{x} \geq \mathbf{0}$ in (P) by a *quadratic regularization* of the linear objective function $\mathbf{c}^T\mathbf{x}$ in the metric defined by $\mathbf{X}_k^{-2}$. The quadratic term in (8.6) can be written as

$$\sum_{j=1}^{n}\left(\frac{x_j - x_j^{(k)}}{x_j^{(k)}}\right)^2, \tag{8.7}$$

so the metric employed is a measure of distance between x_j and $x_j^{(k)}$ taken *relative* to $x_j^{(k)}$, $1 \leq j \leq n$.

The DfP defined by (8.6) has only equality constraints and can be solved explicitly. Again, make the transformation of variables $\mathbf{x} = \mathbf{x}^{(k)} + \mathbf{Z}\mathbf{w}$, where $\mathbf{w} \in R^{n-m}$ and $\mathbf{Z}$ is of full rank and spans the null space of $\mathbf{A}$;

---

[1] To adapt the development in Section 2.3 to the present context, identify $\mathbf{D}_k^+$ with $\mathbf{X}_k^{-2}$ and $\mathbf{D}_k^{1/2}$ with $\mathbf{X}_k^{-1}$.

see, for example, (7.3). The linear constraints hold identically, giving the unconstrained quadratic model

$$\text{minimize } \mathbf{c}^T(\mathbf{Z}^T\mathbf{w}) + \tfrac{1}{2}\mathbf{w}^T(\mathbf{Z}^T\mathbf{X}_k^{-2}\mathbf{Z})\mathbf{w}.$$

By forming the direction to its minimizing point and reverting to the original variables, one again obtains the Dikin affine-scaling direction, namely,

$$\mathbf{d} = -\mathbf{Z}\left(\mathbf{Z}^T\mathbf{X}_k^{-2}\mathbf{Z}\right)^{-1}\mathbf{Z}^T c = -\mathbf{X}_k\mathbf{P}\mathbf{X}_k\mathbf{c}, \tag{8.8}$$

where $\mathbf{P} = \mathbf{I} - \mathbf{X}_k\mathbf{A}^T(\mathbf{A}\mathbf{X}_k^2\mathbf{A}^T)^{-1}\mathbf{A}\mathbf{X}_k$.

8.1.4 Equivalence

The three NC approaches yield the *same* primal affine-scaling search direction. This is a direct consequence of linearity of the objective function $\mathbf{c}^T\mathbf{x}$ (and the choice $\mathcal{H}_k = \mathbf{0}$ in expression (2.19)).

There are two expressions for the primal affine-scaling direction, namely, (7.5) and (7.6). These are called the *range-space* form and the *null-space* form, respectively. The null-space form has advantages for computation, as we have seen in Chapter 7.

The primal affine-scaling direction can be incorporated into an interior-point algorithm along lines considered in Section 7.3. We will return to issues of convergence and implementation of this algorithm in the concluding section of the present chapter.

Because the three Newton–Cauchy approaches to the formulation of affine scaling for linear programming are essentially equivalent, *we will choose the approach that is the most convenient*, namely, the NC quadratic (Dikin) model perspective of Section 8.1.3, in our derivation of dual affine scaling and primal–dual affine scaling that follows.

8.2 Dual Affine Scaling

Let us now consider affine scaling on the dual linear program (D) in (8.1). Assume that $\pi^{(k)}$, $\mathbf{v}^{(k)} > \mathbf{0}$ is a feasible interior point for (D) and let $\mathbf{V}_k = \text{diag}\left[v_1^{(k)}, \ldots, v_n^{(k)}\right]$. In a development completely analogous to (8.6), but now applied to the dual problem (D), define the dual affine-scaling direction as a step from the current iterate $\left(\pi^{(k)}, \mathbf{v}^{(k)}\right)$ to the maximizing point of the following positive definite quadratic approximating model or direction-finding problem (DfP):

$$\begin{aligned} &\text{maximize } \mathbf{b}^T\left(\pi - \pi^{(k)}\right) - \tfrac{1}{2}\left(\mathbf{v} - \mathbf{v}^{(k)}\right)^T\mathbf{V}_k^{-2}\left(\mathbf{v} - \mathbf{v}^{(k)}\right) \\ &\text{s.t. } \mathbf{A}^T\pi + \mathbf{v} - \mathbf{c} = \mathbf{A}^T\left(\pi - \pi^{(k)}\right) + \left(\mathbf{v} - \mathbf{v}^{(k)}\right) = \mathbf{0}. \end{aligned} \tag{8.9}$$

A matrix $\mathbf{Z}$ of full rank, which spans the null space of the matrix defining the linear equality constraints of the dual linear program (D), is

$$\mathbf{Z} = \begin{bmatrix} \mathbf{I} \\ -\mathbf{A}^T \end{bmatrix}.$$

By using $\mathbf{Z}$ to perform a transformation of variables analogous to that of the primal setting, one obtains the dual affine-scaling direction $\mathbf{h} \in R^{m+n}$, whose components correspond to the π and $\mathbf{v}$ variables, as follows:

$$\begin{bmatrix} \mathbf{h}_\pi \\ \mathbf{h}_v \end{bmatrix} = \begin{bmatrix} \mathbf{I} \\ -\mathbf{A}^T \end{bmatrix} (\mathbf{A}\mathbf{V}_k^{-2}\mathbf{A}^T)^{-1}\mathbf{b}. \tag{8.10}$$

This is the *null-space* form of the search direction.

The corresponding range-space form for the dual (D) does *not* exist. We leave the verification of this fact as an exercise for the reader.

8.3 Primal–Dual Affine Scaling

Finally, consider the primal and dual linear programming problems simultaneously. Let $\mathbf{D}_X$ and $\mathbf{D}_V$ be *arbitrary* diagonal positive definite (scaling) matrices. Again employing the NC quadratic or Dikin model perspective, as in Sections 8.1.3 and 8.2, define the following local quadratic approximating models (DfPs) for the primal and dual problems:

$$\begin{aligned} &\text{minimize } \mathbf{c}^T(\mathbf{x} - \mathbf{x}^{(k)}) + \tfrac{1}{2}(\mathbf{x} - \mathbf{x}^{(k)})^T \mathbf{D}_X^{-2} (\mathbf{x} - \mathbf{x}^{(k)}) \\ &\quad \text{s.t. } \mathbf{A}\mathbf{x} - \mathbf{b} = \mathbf{A}(\mathbf{x} - \mathbf{x}^{(k)}) = \mathbf{0}, \end{aligned} \tag{8.11}$$

and

$$\begin{aligned} &\text{maximize } \mathbf{b}^T(\pi - \pi^{(k)}) - \tfrac{1}{2}(\mathbf{v} - \mathbf{v}^{(k)})^T \mathbf{D}_V^{-2} (\mathbf{v} - \mathbf{v}^{(k)}) \\ &\quad \text{s.t. } \mathbf{A}^T\pi + \mathbf{v} - \mathbf{c} = \mathbf{A}^T(\pi - \pi^{(k)}) + (\mathbf{v} - \mathbf{v}^{(k)}) = \mathbf{0}, \end{aligned} \tag{8.12}$$

where, as before, $\mathbf{x}^{(k)}$ and $(\pi^{(k)}, \mathbf{v}^{(k)})$ define feasible interior points of (P) and (D), respectively.

Given a convex quadratic program like (8.11), the program dual to it has a Hessian matrix $\mathbf{D}_X^2$ associated with its slack variables. Thus, the condition $\mathbf{D}_X^2 = \mathbf{D}_V^{-2}$ or, equivalently, $\mathbf{D}_V = \mathbf{D}_X^{-1}$ is a natural one to impose. Another way to justify this condition is that norms that are dual to one another are employed in the two DfPs. Thus, let us consider (8.11) and (8.12) simultaneously, impose the condition $\mathbf{D}_V = \mathbf{D}_X^{-1}$ on them, and require that the optimal primal and dual variables obtained from (8.11) be the same as the corresponding quantities obtained from (8.12); we call this the *self-dual condition.* It results in a *primal–dual affine-scaling* choice for $\mathbf{D}_X$ and $\mathbf{D}_V$, which we will now derive.

The Lagrangian conditions at the optimal solution, say, $\mathbf{x}^*$, of the DfP (8.11), with Lagrange multipliers, say, π^*, are as follows:

$$\begin{aligned} \mathbf{A}\mathbf{x}^* - \mathbf{b} &= \mathbf{0}, \\ \mathbf{A}^T\pi^* - \mathbf{c} - \mathbf{D}_X^{-2}(\mathbf{x}^* - \mathbf{x}^{(k)}) &= \mathbf{0}. \end{aligned} \tag{8.13}$$

Similarly, the Lagrangian conditions at the optimal solution of the DfP (8.12) are as follows, where we use the same notation $\mathbf{x}^*$ and π^* as in (8.13), because we shall seek to satisfy the self-dual condition as defined above:

$$\begin{aligned} \mathbf{A}\mathbf{x}^* - \mathbf{b} &= \mathbf{0}, \\ \mathbf{x}^* + \mathbf{D}_V^{-2}(\mathbf{v}^* - \mathbf{v}^{(k)}) &= \mathbf{0}, \\ \mathbf{A}^T\pi^* + \mathbf{v}^* &= \mathbf{c}. \end{aligned} \tag{8.14}$$

Use the second equation to eliminate $\mathbf{v}^*$ from the third, giving

$$\begin{aligned} \mathbf{A}\mathbf{x}^* - \mathbf{b} &= \mathbf{0}, \\ \mathbf{A}^T\pi^* - \mathbf{D}_V^2\mathbf{x}^* + \mathbf{v}^{(k)} &= \mathbf{c}. \end{aligned} \tag{8.15}$$

Identity between the two sets of optimality conditions (8.13) and (8.15) is ensured by the equation

$$-\mathbf{D}_X^{-2}(\mathbf{x}^* - \mathbf{x}^{(k)}) = -\mathbf{D}_V^2\mathbf{x}^* + \mathbf{v}^{(k)}. \tag{8.16}$$

Using $\mathbf{D}_X = \mathbf{D}_V^{-1}$ then gives

$$\mathbf{D}_X^{-2}\mathbf{x}^{(k)} = \mathbf{v}^{(k)},$$

which is equivalent to

$$\mathbf{D}_X = \mathrm{diag}\left[\sqrt{\left(x_1^{(k)}/v_1^{(k)}\right)}, \ldots, \sqrt{\left(x_n^{(k)}/v_n^{(k)}\right)}\right] = \mathbf{D}_V^{-1}. \tag{8.17}$$

This defines the primal–dual affine-scaling matrix.

The primal–dual affine-scaling *directions* are obtained analogously to the primal and the dual affine-scaling directions with $\mathbf{X}_k$ and $\mathbf{V}_k$ being replaced by $\mathbf{D}_X$ and $\mathbf{D}_V$ and the primal–dual scaling choice (8.17) being made for these two matrices. The primal component of the primal–dual affine-scaling direction is

$$\mathbf{d} = -\mathbf{Z}(\mathbf{Z}^T\mathbf{D}_X^{-2}\mathbf{Z})^{-1}\mathbf{Z}^T\mathbf{c} = -\mathbf{D}_X\mathbf{P}\mathbf{D}_X\mathbf{c}, \tag{8.18}$$

where $\mathbf{P} = \mathbf{I} - \mathbf{D}_X\mathbf{A}^T(\mathbf{A}\mathbf{D}_X^2\mathbf{A}^T)^{-1}\mathbf{A}\mathbf{D}_X$ and $\mathbf{D}_X$ is defined by (8.17).

The dual components of the primal–dual affine scaling direction are

$$\begin{bmatrix} \mathbf{h}_\pi \\ \mathbf{h}_v \end{bmatrix} = \begin{bmatrix} \mathbf{I} \\ -\mathbf{A}^T \end{bmatrix} (\mathbf{A}\mathbf{D}_V^{-2}\mathbf{A}^T)^{-1}\mathbf{b} = \begin{bmatrix} \mathbf{I} \\ -\mathbf{A}^T \end{bmatrix} (\mathbf{A}\mathbf{D}_X^2\mathbf{A}^T)^{-1}\mathbf{b} \tag{8.19}$$

with $\mathbf{D}_V$ and $\mathbf{D}_X$ defined by (8.17).

Note that the primal and the dual directions can be obtained from the factorization of the *same* matrix $(\mathbf{A}\mathbf{D}_X^2\mathbf{A}^T)$.

8.4 The Central Path

Let us require that the components of the primal and dual feasible interior points of (P) and (D) satisfy the following condition:

$$x_j^{(k)} v_j^{(k)} = \mu, \quad j = 1, \ldots, n, \tag{8.20}$$

where μ is any positive scalar. Then it is immediately evident that the primal–dual affine-scaling matrix $\mathbf{D}_X$ in (8.17) differs by a positive scalar multiple from the primal affine-scaling matrix $\mathbf{X}_k$, and that the primal–dual affine-scaling matrix $\mathbf{D}_V$ differs by a positive scalar multiple from the dual affine-scaling matrix $\mathbf{V}_k$. Furthermore, the associated primal and dual components of the search directions using primal–dual affine-scaling of Section 8.3 are *parallel* to the primal affine- and dual affine-scaling directions, respectively, of Sections 8.1 and 8.2. Thus, it is natural to investigate the restriction (8.20) on the feasible interior points $\mathbf{x}^{(k)}$ and $\left(\pi^{(k)}, \mathbf{v}^{(k)}\right)$. Such points must simultaneously satisfy the following set of conditions:

$$\begin{aligned} \mathbf{A}\mathbf{x}^{(k)} &= \mathbf{b}, \\ \mathbf{A}^T\pi^{(k)} + \mathbf{v}^{(k)} &= \mathbf{c}, \\ \mathbf{X}_k\mathbf{V}_k\mathbf{e} &= \mu\mathbf{e}, \\ \mathbf{x}^{(k)} > \mathbf{0}, \quad \mathbf{v}^{(k)} &> \mathbf{0}, \ \mu > 0, \end{aligned} \tag{8.21}$$

where $\mathbf{e}$ is an n-vector of 1's.

We will see later, in Chapters 10 and 11, that (8.21) has a *unique* solution $\left(\mathbf{x}^{(k)}, \pi^{(k)}, \mathbf{v}^{(k)}\right)$ for each value of $\mu > 0$, and as μ is varied continuously, the solutions define a fundamental object of linear programming known as the *central path*. Here our objective is simply to observe that the set of conditions that characterize the central path arises in a very natural way from the NC-based line of development of primal, dual, and primal–dual affine-scaling matrices in the present chapter and from the relationships among their associated search directions.

8.5 Convergence and Implementation

From the definitions of the primal, dual, and primal–dual affine-scaling directions of Sections 8.1–8.3, it is evident, *using the null-space expressions* given in (8.5), (8.10), (8.18), and (8.19), that the primal directions are directions of *descent* for the primal objective function $\mathbf{c}^T\mathbf{x}$, and that the dual directions are directions of *ascent* for the dual objective functions $\mathbf{b}^T\pi$.

Once a search direction is obtained, an improving iterate can be found along it, subject to remaining in the positive orthant with respect to the $\mathbf{x}$ and/or $\mathbf{v}$ variables. Primal, dual, or primal–dual affine-scaling algorithms can then be devised along lines that are completely analogous to the primal affine-scaling algorithm described in Chapter 7. They can utilize either the range-space form (when it exists) or the null-space form of the search directions.

The proof of convergence of any affine-scaling algorithm is a challenging task. Convergence under nondegeneracy assumptions was originally proved by Dikin [1967], [1974]. After affine scaling was independently discovered as an affine variant of Karmarkar's projective-scaling algorithm by many different researchers, proofs of convergence were obtained again by Vanderbei, Meketron, and Freedman [1986], Barnes [1986], and others, and, for the case of degenerate problems, by Tsuchiya and Muramatsu [1995]. For a detailed discussion, see the comprehensive monograph of Saigal [1995] and the references cited therein.

Convergence of a *pure* primal–dual affine-scaling algorithm, with *independent* step lengths in the primal and dual spaces, has not been established to date. However, a variant that can activate the restriction (8.20), in order to link the primal and dual iterates under certain rare circumstances, has been shown to be convergent. For details, see Kojima, Megiddo, Noma, and Yoshise [1991].

Pure primal or dual affine-scaling algorithms are not known to have polynomial complexity, but variants that are permitted to augment the search direction by a "centering component" have been shown to exhibit this property. Details are given, for example, in Barnes, Chopra, and Jensen [1987] and Saigal [1995].

The practical implementation of affine-scaling algorithms has been studied extensively. See, for example, Adler, Karmarkar, Resende, and Viega [1989] for realistic numerical experiments with a dual affine-scaling algorithm, or see Kim and Nazareth [1994] for a study of a null-space version of the primal affine-scaling algorithm. As observed by Dantzig and Thapa [1997] and others, *hybrid* implementations that combine simplex and affine-scaling techniques within a mathematical programming system are likely to be more effective than either of their individual component algorithms when used to solve the large-scale linear programming problems that arise in practice.

8.6 Notes

Sections 8.1–8.3: This material is derived from Nazareth [1994c].

9

Diagonal Metrics and the QC Method

The previous two chapters considered Cauchy techniques based on diagonally scaled projected steepest descent for solving a linear program. We now turn to diagonally scaled Cauchy-type techniques for solving unconstrained *nonlinear* minimization problems.

By way of motivation, consider the following observation of Bertsekas [1999] concerning the minimization of a nonlinear objective function with a "difficult" topography:[1]

> Generally, there is a tendency to think that difficult problems should be addressed with sophisticated methods, such as Newton-like methods. This is often true, particularly for problems with nonsingular local minima that are poorly conditioned. However, it is important to realize that *often the reverse is true*, namely, that for problems with "difficult" cost functions and singular local minima, it is best to use *simple methods such as (perhaps diagonally scaled) steepest descent* with simple step size rules such as constant or diminishing step size. The reason is that methods that use sophisticated descent directions and step size rules often rely on assumptions that are likely to be violated in difficult problems. We also note that for difficult problems, it may be helpful to supplement the steepest descent method with features that allow it to better deal with multiple

[1]Italics in this quotation are ours.

local minima and peculiarities of the cost function. An often useful modification is the heavy ball method....

The gradient vector used in the foregoing Cauchy-type methods can be obtained in one of three principal ways:

- by using analytic expressions for partial derivatives;
- by using automatic differentiation;
- by taking n finite differences of function values over numerically infinitesmal steps along, for example, the coordinate axes.

The first two options are often not available and the third can be prohibitively expensive when the problem dimension is large and the computation of each function value is itself expensive. Thus, we set out to explore "simple" Cauchy-type techniques, akin to those mentioned in the foregoing quotation, that, additionally, do *not* require a *full* gradient vector. We will attach the name *quasi-Cauchy* (QC) to the resulting approach.

A metric-based quasi-Cauchy algorithm for unconstrained minimization makes assumptions about the objective function that are similar to those required for the steepest descent algorithm, and it has the following general characteristics:

1. Instead of requiring a full gradient vector, it develops its search direction of *descent* at $\mathbf{x}_k$ from a relatively small set of directional derivatives derived from finite differences of function values at the current iterate.[2]
2. It employs a *line search*, or possibly a simpler step-size procedure, for example, Armijo–Goldstein (see Kelley [1999] and references therein), using only function values and/or directional derivatives obtained by finite differences of function values, in order to find an improved iterate, say $\mathbf{x}_{k+1}$, where $f(\mathbf{x}_{k+1}) < f(\mathbf{x}_k)$.
3. It utilizes a metric defined by a positive definite *diagonal* matrix.

We formulate an *illustrative* QC algorithm based on particular choices for each of the foregoing three items, and we give the results of some numerical experimentation with this algorithm. Refinements of each of its main ingredients are then discussed, in particular, the technique of *QC diagonal updating*.

Quasi-Cauchy algorithms occupy the *middle ground* between diagonally scaled steepest descent and cyclic coordinate descent algorithms. They rep-

[2]This chapter is closely integrated with the Newton–Cauchy framework of Chapter 2 and thus uses our primary notation for iterates; see the comments on notation on page xvii.

resent a very natural extension of the Newton–Cauchy framework of Chapter 2, which is given in the concluding section of this chapter.

9.1 A Quasi-Cauchy Algorithm

Let $\mathbf{D}_k$ be a diagonal matrix[3] with positive diagonal elements. At any point, say $\mathbf{x}_k$, with gradient vector $\mathbf{g}_k$, a Cauchy algorithm uses the direction of steepest descent in the metric defined by $\mathbf{D}_k$, namely, the direction $-\mathbf{D}_k^{-1}\mathbf{g}_k$.

Let us instead form an approximation termed the *quasi-gradient*, say $\overline{\mathbf{g}}_k$, to the gradient $\mathbf{g}_k$, in a way that ensures that the direction $\mathbf{d}_k = -\mathbf{D}_k^{-1}\overline{\mathbf{g}}_k$ is a direction of descent. We then conduct a line search along $\mathbf{d}_k$, update $\mathbf{D}_k$ to a diagonal matrix $\mathbf{D}_{k+1}$ so as to incorporate curvature information gathered from the line search, and repeat the process.

9.1.1 Descent Direction

Within the foregoing context, the technique for defining a search direction at $\mathbf{x}_k$ will be based on a simple result as follows:

Proposition 9.1: Let $U = \{\mathbf{u}_1, \ldots, \mathbf{u}_l\}$ be a set of l nonzero directions at the point $\mathbf{x}_k$ with gradient vector $\mathbf{g}_k$, and let $\mathbf{D}_k$ be a positive definite diagonal matrix. Let $\overline{\mathbf{g}}_k$ be any nonzero vector that approximates $\mathbf{g}_k$ in the sense that

$$\overline{\mathbf{g}}_k^T \mathbf{u}_i = \mathbf{g}_k^T \mathbf{u}_i, \quad 1 \leq i \leq l. \tag{9.1}$$

Suppose that $\mathbf{d}_k \equiv -\mathbf{D}_k^{-1}\overline{\mathbf{g}}_k$ is linearly dependent on $\mathbf{u}_1, \ldots, \mathbf{u}_l$. Then $\mathbf{d}_k$ must be a direction of descent.

Proof: From (9.1), $(\overline{\mathbf{g}}_k - \mathbf{g}_k)^T \mathbf{u}_i = 0$, $i = 1, \ldots, l$. If $\mathbf{d}_k$ is linearly dependent on $\mathbf{u}_1, \ldots, \mathbf{u}_l$, then $(\overline{\mathbf{g}}_k - \mathbf{g}_k)^T \mathbf{d}_k = 0$. Thus $\mathbf{g}_k^T \mathbf{d}_k = -\overline{\mathbf{g}}_k^T \mathbf{D}_k^{-1} \overline{\mathbf{g}}_k < 0$. □

Take $2 \leq l \ll n$, and group the n coordinate vectors into mutually exclusive blocks, each containing $(l - 1)$ vectors, These blocks, say L in number, are used, one at a time, in a cyclic sequence. At the current iterate $\mathbf{x}_k$, the directional derivative along each vector in the block that comes up next in the sequence (the "current block") is estimated by finite differences in the standard way using, for example, an $O(macheps^{1/2})$ step from $\mathbf{x}_k$ along the vector, where *macheps* denotes the machine-precision constant. In addition, as we will see in the next subsection, the line search along the

[3]The superscript +, which was used in Chapter 2 to explicitly distinguish positive definite matrices, is no longer needed here, and henceforth it will be omitted.

previous search direction, say $\mathbf{d}_{k-1}$, that found the current iterate $\mathbf{x}_k$, will exit with the directional derivative $\mathbf{g}_k^T\mathbf{d}_{k-1}$ at $\mathbf{x}_k$ along $\mathbf{d}_{k-1}$. Thus, the set of l vectors consisting of the $(l-1)$ vectors in the current block and the vector $\mathbf{d}_{k-1}$ will constitute the vectors $\mathbf{u}_1, \ldots, \mathbf{u}_l$ of Proposition 9.1. These vectors have known directional derivatives at $\mathbf{x}_k$.

The quantity $\overline{\mathbf{g}}_k$, the quasi-gradient, is then found by minimizing the Euclidean norm of $\overline{\mathbf{g}}_k$ subject to satisfying the constraints (9.1). This is a low-dimensional linear least-squares problem that can be solved very efficiently.

Finally, the search direction at $\mathbf{x}_k$ is defined by $\mathbf{d}_k = -\mathbf{D}_k^{-1}\overline{\mathbf{g}}_k$. The choice of the matrix $\mathbf{D}_k$ defining the metric is considered in Section 9.1.3. The directional derivative $\mathbf{g}_k^T\mathbf{d}_k$ along $\mathbf{d}_k$ is obtained by a finite difference. If its sign establishes that $\mathbf{d}_k$ is a direction of descent, then proceed to the line search. If not, then the above proposition tells us that $\mathbf{d}_k$ must be linearly *independent* of the set U. Add $\mathbf{d}_k$ and its directional derivative to the set of directions that are used to obtain the quasi-gradient; i.e., $l \leftarrow l+1$, $U \leftarrow U \cup \mathbf{d}_k$, and repeat the procedure to obtain a new quasi-gradient $\overline{\mathbf{g}}_k$, search direction $\mathbf{d}_k = -\mathbf{D}_k^{-1}\overline{\mathbf{g}}_k$, and the new directional derivative. Obviously, it must terminate in a finite number of steps with a direction of descent.

9.1.2 Line Search

The direction $\mathbf{d}_k$ and its directional derivative $\mathbf{g}_k^T\mathbf{d}_k$ are used to initiate a standard line search based, for example, on fitting a cubic polynomial to function and directional derivative information. An example of such a line search is given in Chapter 5 for the case where directional derivatives are obtained from gradient vectors. Now, instead, when the line search defines a fresh candidate point, say $\mathbf{x}_{k+1}$, along the line passing through $\mathbf{x}_k$ parallel to $\mathbf{d}_k$ and requires the directional derivative, say $\mathbf{g}_{k+1}^T\mathbf{d}_k$, at $\mathbf{x}_{k+1}$, this information is *again obtained by a finite-difference operation.* The line search can be terminated, for example, when the Wolfe conditions are satisfied or, more simply, when $|\mathbf{g}_{k+1}^T\mathbf{d}_k|/|\mathbf{g}_k^T\mathbf{d}_k| \leq \mathrm{ACC}$, where $\mathrm{ACC} \in (0,1)$. These conditions imply $\left(\mathbf{g}_{k+1}^T\mathbf{d}_k - \mathbf{g}_k^T\mathbf{d}_k\right) > 0$. Note that several intermediate points, each denoted by $\mathbf{x}_{k+1}$ in the foregoing discussion, may be generated before the line search termination criterion is satisfied; i.e., several directional derivatives may be used if a high accuracy is imposed on the line search, e.g., ACC $= 0.1$. With a larger value for the accuracy parameter ACC, for example, 0.9, the line search will usually exit with very few function evaluations.

9.1.3 Diagonal Metric

The metric used to define the search direction $\mathbf{d}_k$ at $\mathbf{x}_k$ will be defined by a diagonal matrix, say $\mathbf{D}_k$, with positive diagonal elements. The line search along $\mathbf{d}_k = -\mathbf{D}_k^{-1}\mathbf{g}_k$ is entered at $\mathbf{x}_k$ with slope information $\mathbf{g}_k^T\mathbf{d}_k$ and exits at $\mathbf{x}_{k+1}$ with slope information $\mathbf{g}_{k+1}^T\mathbf{d}_k$. Define $\mathbf{s}_k = \mathbf{x}_{k+1} - \mathbf{x}_k \neq \mathbf{0}$ and $\mathbf{y}_k = \mathbf{g}_{k+1} - \mathbf{g}_k$, and note that $\mathbf{y}_k$ is *not* available to the algorithm. Let $a_k = \mathbf{s}_k^T\mathbf{D}_k\mathbf{s}_k > 0$ and $b_k = \mathbf{y}_k^T\mathbf{s}_k$. The latter quantity is available to the algorithm by computing $b_k = (\mathbf{g}_{k+1}^T\mathbf{d}_k - \mathbf{g}_k^T\mathbf{d}_k)(\|\mathbf{s}_k\|/\|\mathbf{d}_k\|) > 0$.

The new curvature information derived from the line search is incorporated into the updated diagonal matrix, say $\mathbf{D}_{k+1}$, which defines the metric at $\mathbf{x}_{k+1}$ in the following way. Let $\mathbf{M}_{k+1}$ be the matrix obtained by updating $\mathbf{D}_k$ by a matrix of rank one, namely,

$$\mathbf{M}_{k+1} = \mathbf{D}_k + \frac{(b_k - a_k)}{a_k^2}\mathbf{D}_k\mathbf{s}_k\mathbf{s}_k^T\mathbf{D}_k, \qquad (9.2)$$

and let $\mathbf{D}_{k+1}$ be the diagonal matrix with $(\mathbf{D}_{k+1})_{ii} = (\mathbf{M}_{k+1})_{ii}$, $i = 1, \ldots, n$. This matrix can be computed directly as follows:

$$(\mathbf{D}_{k+1})_{ii} = (\mathbf{D}_k)_{ii} + \frac{(b_k - a_k)}{a_k^2}(\mathbf{s}_k)_i^2(\mathbf{D}_k)_{ii}^2.$$

Note that directional derivative information from the computation of the search direction and from the line search is being *fully* used, and that only "hard" derivative information, as constrasted with "soft" quasi-gradient estimates of derivative information, is used to update the metric.

The motivation for using the matrix $\mathbf{M}_{k+1}$ is that it satisfies a weakened form[4] of the quasi-Newton relation (2.5) of Chapter 2, namely,

$$\mathbf{s}_k^T\mathbf{M}_{k+1}\mathbf{s}_k = b_k \equiv \mathbf{y}_k^T\mathbf{s}_k. \qquad (9.3)$$

Under the conditions $a_k = \mathbf{s}_k^T\mathbf{D}_k\mathbf{s}_k > 0$ and $b_k > 0$, it is easy to show that $\mathbf{M}_{k+1}$ defined by (9.2) is positive definite. (We leave this as an exercise for the interested reader.) Thus $\mathbf{D}_{k+1}$ must have positive diagonal elements. Note, however, that the foregoing matrix $\mathbf{D}_{k+1}$, in general, does *not* satisfy the weak quasi-Newton relation (9.3).

9.1.4 Numerical Illustration

The foregoing algorithm was directly encoded in Matlab and run with the setting ACC = 0.5 in the line search, the choice $(l - 1) = 5$ in the QC

[4]This so-called *weak quasi-Newton* or secant condition (9.3) is due to Dennis and Wolkowicz [1993]. There is a subtle but important difference in our usage here, because Dennis and Wolkowicz [1993] assume that the vector $\mathbf{y}_k$ is available—their setting for usage is quasi-Newton—and much of their theoretical development involves updates defined in terms of $\mathbf{y}_k$ or the quantity $c_k = \mathbf{y}_k^T\mathbf{D}_k^{-1}\mathbf{y}_k$. An exception is the update labeled (4.10) in Dennis and Wolkowicz [1993], which corresponds to our expression (9.2) when b_k is computed from directional derivatives rather than gradients.

TABLE 9.1. Performance of a QC algorithm.

Quasi-Cauchy		Cauchy		Cyclic Coord.	
507	5.62	502	190.80	511	27.77
600	3.89	600	121.26	601	22.67
701	0.051	698	92.04	727	0.49
806	6.84e-4	796	60.23	817	0.35
900	4.81e-5	896	39.26	908	0.34
1003	8.71e-7	1026	29.23	1020	0.013

method, and again with the choice $(l-1) = n$, which, in effect, yields the Cauchy method with the diagonal scaling (or preconditioning) of Section 9.1.3. In addition, the cyclic coordinate descent method with a line search was implemented in Matlab for purposes of comparison. The three algorithms were run on a classical test problem, the trignometric function of Fletcher and Powell [1963], using its standard starting point and the choice of problem dimension $n = 30$; for details of the test problem, see Moré, Garbow, and Hillstrom [1981]. The optimal value for its objective function is 0.

For each of the three algorithms, Table 9.1 gives a list of six function *counts*; i.e., the number of calls to the function evaluation routine, and corresponding objective function *values*. These quantities were recorded periodically, the first entry for an algorithm being recorded at the iterate where the function count exceeded 500, and thereafter the count was recorded in approximate multiples of 100 until the number of function calls exceeded 1000. Thus, for example, in line 3 of the table, the Quasi-Cauchy algorithm attained the objective function value 0.051, and at that point it had made 701 calls to the function evaluation routine. In contrast, the Cauchy algorithm required almost the same number of function calls, namely 698, to reduce the objective function value to only 92.04.

It is interesting to observe how much more progress is made by the QC algorithm in comparison to the other two algorithms. Note also the greater efficiency of the cyclic coordinate descent algorithm in comparison with the diagonally preconditioned Cauchy algorithm.

9.2 The QC Method

Let us consider refinements of all the three main ingredients of the illustrative QC algorithm.

9.2.1 *Quasi-Gradient*

Several improvements are possible, for example, the following:

1. If very little progress is made in the line search along a direction, say $\mathbf{d}_{k-1}$, then the set of vectors in the previous block of vectors used at $\mathbf{x}_{k-1}$ and their associated directional derivatives could be included in the set $\mathbf{u}_1, \ldots, \mathbf{u}_l$ used at $\mathbf{x}_k$. Clearly, this will eventually yield a good approximation to the gradient if performed repeatedly.
2. Additionally or alternatively, the quasi-gradient $\overline{\mathbf{g}}_k$ can be chosen to minimize the Euclidean norm of $\overline{\mathbf{g}}_k - \overline{\mathbf{g}}_{k-1}$, where $\overline{\mathbf{g}}_{k-1}$ is the quasi-gradient estimate at $\mathbf{x}_{k-1}$. It is easy to construct an example where this would result in a nondescent direction at $\mathbf{x}_k$, bringing Proposition 9.1 directly into play.
3. A basis other than the coordinate vectors can be employed. In addition, one or more randomly chosen directions, with directional derivatives estimated by finite differences along them, could be added to the set used to estimate the quasi-gradient at each iterate. Introducing some randomization into the algorithm could be very useful.
4. If the set of vectors used to estimate a quasi-gradient at an iterate does not yield a direction of descent and must be augmented, then it is possible to use well-known computational linear algebra techniques to solve the least-squares problem and obtain the new quasi-gradient very efficiently.

9.2.2 *Line Search*

A line search along the direction defined by iterates at the beginning and end of a full cycle of L blocks, as in some pattern-search methods, could be a useful acceleration step. This would again use the routine described in Section 9.1.2. The initial sign of the directional derivative would determine the direction of the ray (half-line) along which the search is conducted.

Simple line-search routines, as mentioned in the quotation at the beginning of this chapter, could be substituted, and the diagonal update performed only when the new iterate satisfies the condition $b_k > 0$.

9.2.3 *Diagonal Updating*

We have noted that the matrix $\mathbf{M}_{k+1}$ given by (9.2) satisfies the weak quasi-Newton relation (9.3), but the matrix $\mathbf{D}_{k+1}$, obtained from the diagonal elements of $\mathbf{M}_{k+1}$ and used to define the metric, does not. Let us now consider refinements of the scaling techniques where the diagonal matrix that is used to define the metric in the QC algorithm is required to satisfy a relation like (9.3).

We define the *quasi-Cauchy (QC) relation* to be the weak quasi-Newton relation (9.3) with the added restriction that the matrix $\mathbf{M}_{k+1}$ must be a *diagonal* matrix. Thus, the quasi-Cauchy relation is as follows:

$$\mathbf{s}_k^T \mathbf{D}_{k+1} \mathbf{s}_k = b_k \equiv \left(\mathbf{g}_{k+1}^T \mathbf{s}_k - \mathbf{g}_k^T \mathbf{s}_k\right). \tag{9.4}$$

As in the discussion in Sections 9.1.2 and 9.1.3, the quantity $b_k = \mathbf{y}_k^T \mathbf{s}_k$ in the right-hand side of the QC relation can be obtained by directional derivative differences along $\mathbf{s}_k$. Thus explicit knowledge of full gradient vectors is not needed.

As an example, consider the well-known scaling matrix of Oren and Luenberger [1974]:

$$\mathbf{D}_{k+1} = \left(\mathbf{s}_k^T \mathbf{y}_k / \mathbf{s}_k^T \mathbf{s}_k\right) \mathbf{I},$$

where $\mathbf{I}$ is the identity matrix. It is interesting to note that this is precisely the unique matrix that would be obtained under the requirement that the QC relation be satisfied and with the added restriction that the diagonal matrix must be a scalar multiple of the identity matrix. In other words, the diagonal elements of the matrix $\mathbf{D}_{k+1}$ are required to be equal, and the associated quadratic approximating model has contours that are hyperspheres. Note also that Oren–Luenberger scaling is a rank-n change to the matrix $\mathbf{I}$. The scaling matrices derived from the QC relation can be viewed as a natural *generalization* of Oren–Luenberger scaling.

Thus, suppose $\mathbf{D}_k$ is a given positive definite diagonal matrix and $\mathbf{D}_{k+1}$, which is also diagonal, is the updated version of $\mathbf{D}_k$. Let us require that the updated $\mathbf{D}_{k+1}$ satisfy the QC relation and that the deviation between $\mathbf{D}_k$ and $\mathbf{D}_{k+1}$ be minimized under some variational principle. (Here we will use only the Frobenius matrix norm to measure the deviation.) We would like the derived update to preserve positive definiteness in a natural way; i.e., we seek well-posed variational problems such that the solution $\mathbf{D}_{k+1}$, through the diagonal updating procedure, incorporates available curvature information from the step and gradient changes, as well as that contained in $\mathbf{D}_k$. As noted earlier, a diagonal matrix uses the same computer storage as a vector, so only $O(n)$ storage is required. Thus, the resulting update will have potential use in algorithms where storage is at a premium. An obvious usage would be to recondition or rescale Cauchy's steepest descent direction, which accounts for our earlier choice of terminology.

A variety of variational problems can be formulated for which (9.4) is a constraint that must be satisfied, and they lead to different updates of $\mathbf{D}_k$ to $\mathbf{D}_{k+1}$, some of which preserve positive definiteness when $\mathbf{D}_k$ is positive definite. Two examples are given below:

- The analogue of the variational principle used to define the so-called Powell–Symmetric–Broyden or PSB update in the quasi-Newton setting (Dennis and Schnabel [1983]) yields the variational problem

$$\begin{aligned} &\text{minimize } \|\mathbf{D}_{k+1} - \mathbf{D}_k\|_F \\ &\text{s.t. } \mathbf{s}_k^T \mathbf{D}_{k+1} \mathbf{s}_k = b_k, \end{aligned} \tag{9.5}$$

where $\mathbf{s}_k \neq \mathbf{0}$, $b_k = \mathbf{y}_k^T \mathbf{s}_k > 0$, and F denotes the Frobenius norm.

- The analogue of the variational principle used to derive the BFGS update in the quasi-Newton setting of Chapter 2 yields a variational problem that updates the square-root or Cholesky factor $\mathbf{D}_k^{1/2}$ to $\mathbf{D}_{k+1}^{1/2}$ as follows:

$$\mathbf{D}_{k+1}^{1/2} = \mathbf{D}_k^{1/2} + \Omega_k,$$

where Ω_k is a diagonal matrix chosen to

$$\begin{aligned} &\text{minimize } \|\Omega_k\|_F \\ &\text{s.t. } \mathbf{s}_k^T \left(\mathbf{D}_k^{1/2} + \Omega_k\right)^2 \mathbf{s}_k = b_k, \end{aligned} \tag{9.6}$$

where $\mathbf{s}_k \neq \mathbf{0}$, $b_k = \mathbf{y}_k^T \mathbf{s}_k > 0$, and again F denotes the Frobenius norm.

The foregoing variational problems are studied, in detail, in Zhu [1997] and Zhu, Nazareth, and Wolkowicz [1999], *within the setting where gradients are available*. In particular, analogously to the full matrix case in standard QN updating, it is shown that (9.6) has a solution for which $\mathbf{D}_{k+1}$ is necessarily positive definite. In contrast, the solution of (9.5) does not have this desirable property and thus cannot always be used to define a metric. (It can be used in a model-based algorithm if desired; see Section 2.2.) These and other results obtained in the references just cited can be directly transported to the present quasi-Cauchy context, wherein all information is obtained via directional derivatives.

Like Oren–Luenberger scaling, the quasi-Cauchy updates obtained from the foregoing variational problems *involve a change of rank n* to the diagonal matrix that defines the metric. In contrast, a quasi-Newton update, say $\mathbf{M}_{k+1}$, generally differs by a matrix of *low rank*, usually one or two, from the prior Hessian approximation matrix $\mathbf{M}_k$.

9.3 Extension of the NC Framework

Observe in the QC algorithm of Section 9.1 that one obtains the cyclic coordinate descent algorithm (CYCD) under the following conditions:

1. $l - 1 = 1$.
2. The latest search direction is *not* included in the set U.

If the latest direction is included in U, the QC algorithm resembles a heavy-ball extension of CYCD; see Section 5.2.3. If a full complement of n di-

Model-Based	Metric-Based
Newton (N)	Modified-Newton (MN)
SR1	BFGS
L-SR1; L-RH-N; L-RH-SR1	L-BFGS; L-RH-MN; L-RH-BFGS
Modified-Cauchy (MC)	Cauchy (C)
UObyQA	QC
"Sampling"	CYCD

FIGURE 9.1 Extended NC framework.

rections defines the set U, then the QC algorithm reverts to a Cauchy algorithm. Thus, generically speaking, quasi-Cauchy (QC) is the middle ground between Cauchy and cyclic coordinate descent (CYCD), just as quasi-Newton (QN) is the middle ground between the Newton and Cauchy approaches.

The QC approach broadens, in an attractive way, the Newton/Cauchy framework of Chapter 2. This is depicted in the metric-based, right-hand, half of Figure 9.1.

The convergence analysis of the QC method and its connections to and contrasts with Gauss–Seidel iteration (when f is a convex quadratic) and block cyclic coordinate descent are well worth further exploration. In this regard, we should also mention the *implicit filtering* algorithms studied, for example, by Carter [1993] and Kelley [1999], which employ inexact gradient vectors. In an implicit filtering algorithm, directional derivative information is not exact in *any* direction, and a search direction is not guaranteed to be a direction of descent. The number of iterations on sample problems has been observed to grow exponentially with a measure of inaccuracy in gradient evaluation (see Carter [1993]). In contrast, a QC algorithm uses *exact* derivative information (up to truncation error in a finite difference estimate) in a subspace of dimension $\ll n$ and always obtains a direction of descent. This could render the QC approach potentially more effective than implicit filtering. The relative merits of the two approaches are worth further investigation.

Powell [2000] has recently proposed a method—unconstrained optimization by quadratic approximation—where both gradient vectors and Hessian matrices are simultaneously estimated from function values at well-separated points. This technique, denoted by UObyQA, is a *model-based counterpart* of the QC approach. It is also a stepping stone to algorithms based on *sampling* the objective function according to some geometric pattern, for example, algorithm NM-GS of Chapter 6 based on a simplex. Other algorithms of this type[5] are pattern search and multidirectional search. For an overview, see Kelley [1999] and references cited therein. The foregoing methods are depicted in the model-based, left-hand, half of Figure 9.1.

The extended part of the NC framework of Figure 9.1 is currently an area of active research in algorithmic optimization.

9.4 Notes

Sections 9.1–9.2: The material in these sections is derived from Nazareth [1995b] and Nazareth [1995c].

[5]Methods in this general area also include simulated annealing, genetic algorithms, and tabu search, but these are used primarily to solve the *global* minimization problem. For a good overview and unification under the rubric of probabilistic search techniques, see Fox [1993]. See also Chapter 14, in particular, Section 14.1.2.

Part IV

Linear Programming Post-Karmarkar

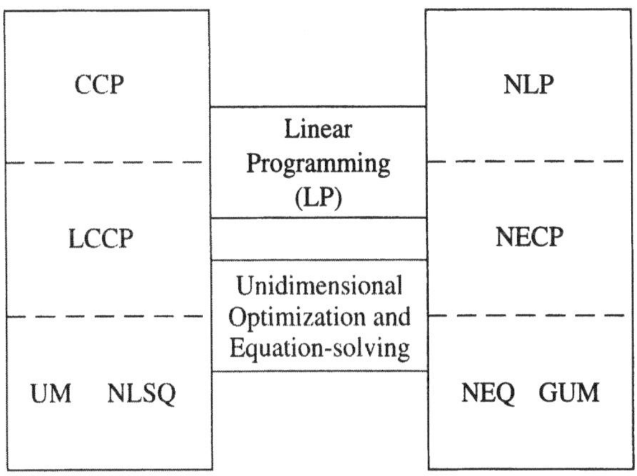

10

LP from the Euler–Newton Perspective

We consider again the primal (P) and dual (D) linear programs (8.1) of Chapter 8. The assumptions stated after expressions (8.1) ensure that optimal solutions of the primal, say $\mathbf{x}^*$, and dual, say $(\pi^*, \mathbf{v}^*)$, simultaneously exist and satisfy primal feasibility, dual feasibility, and complementary slackness; i.e., they satisfy the well-known Karush–Kuhn–Tucker (KKT) optimality conditions:

$$\begin{aligned} \mathbf{Ax} &= \mathbf{b}, \\ \mathbf{A}^T\pi + \mathbf{v} &= \mathbf{c}, \\ \mathbf{XVe} &= \mathbf{0}, \end{aligned} \tag{10.1}$$

and $\mathbf{x} \in R^n_+$, $\mathbf{v} \in R^n_+$. In the third set of equations,

$$\mathbf{X} = \mathrm{diag}[x_1, \ldots, x_n], \qquad \mathbf{V} = \mathrm{diag}[v_1, \ldots, v_n],$$

and $\mathbf{e}$ is an n-vector of 1's. The system (10.1) represents a set of $(m + 2n)$ linear and nonlinear equations in $(m+2n)$ variables $\mathbf{w} = (\mathbf{x}, \pi, \mathbf{v})$. Feasible solutions of the linear programs must also satisfy the nonnegativity bounds on $\mathbf{x}$ and $\mathbf{v}$.

The optimal solutions of (P) and (D) are unique under assumptions of nondegeneracy. Such assumptions are not needed here, but they can be made, if desired, for convenience of discussion.

10.1 The Parameterized Homotopy System

We build on the fundamental EN approach of Section 3.2 and the global homotopy (3.10), now applied to the system (10.1).

10.1.1 Interior Paths

Given initial interior vectors $\mathbf{x}^{(0)} > \mathbf{0}$, $\pi^{(0)}$, and $\mathbf{v}^{(0)} > \mathbf{0}$, which are permitted to be infeasible; i.e., they need not satisfy the equality constraints of (P) or (D), define the global homotopy parameterization of the equations (10.1) as follows:

$$\begin{aligned} \mathbf{A}\mathbf{x} &= \mathbf{b} + \mu \mathbf{q}_{x^{(0)}}, \\ \mathbf{A}^T \pi + \mathbf{v} &= \mathbf{c} + \mu \mathbf{q}_{v^{(0)}}, \\ \mathbf{X}\mathbf{V}\mathbf{e} &= \mu \mathbf{X}_0 \mathbf{V}_0 \mathbf{e}, \end{aligned} \tag{10.2}$$

where $\mathbf{x}, \mathbf{v} \in R^n_+$ and $\mu \in [0, 1]$ is a parameter. In the foregoing equations (10.2), $\mathbf{q}_{x^{(0)}} = \mathbf{A}\mathbf{x}^{(0)} - \mathbf{b}$, $\mathbf{q}_{v^{(0)}} = \mathbf{A}^T\pi^{(0)} + \mathbf{v}^{(0)} - \mathbf{c}$, $\mathbf{X}_0 = \mathrm{diag}\left[x_1^{(0)}, \ldots, x_n^{(0)}\right]$, and $\mathbf{V}_0 = \mathrm{diag}\left[v_1^{(0)}, \ldots, v_n^{(0)}\right]$. When $\mu = 1$, a solution of (10.2) is obviously the given initial point $\mathbf{w}^{(0)} = \left(\mathbf{x}^{(0)}, \pi^{(0)}, \mathbf{v}^{(0)}\right)$, and when $\mu = 0$, a solution of (10.2) with $\mathbf{x}, \mathbf{v} \in R^n_+$ is an optimal solution of (P) and (D) in (8.1), say $\mathbf{w}^* = (\mathbf{x}^*, \pi^*, \mathbf{v}^*)$. For each $\mu > 0$, a *unique* solution of (10.2) exists that lies in the nonnegative orthant with respect to the $\mathbf{x}$ and $\mathbf{v}$ variables. A convenient proof is given in the next chapter—see the discussion immediately following expression (11.4)—and an alternative proof is outlined in the paragraphs following expression (10.4) below.

The set D of Section 3.2.1 is now given by vectors $\mathbf{w} = (\mathbf{x}, \pi, \mathbf{v})$ with $\mathbf{x} \in R^n_+$, $\pi \in R^m$, and $\mathbf{v} \in R^n_+$. Note that D is closed but unbounded; i.e., it is not compact. When $\mathbf{x}$ and $\mathbf{v}$ have strictly positive components, then $\mathbf{w}$ is in D^0, the interior of D. Otherwise, it lies on the boundary ∂D. In addition, we define the following quantities for the homotopy system (10.2):

1. For each $\mu > 0$, let us denote the unique solution of (10.2) by $\mathbf{w}(\mu) = (\mathbf{x}(\mu), \pi(\mu), \mathbf{v}(\mu))$. Note, in particular, that $\mathbf{w}(1) = \mathbf{w}^{(0)}$ and $\mathbf{w}(0) = \mathbf{w}^*$.
2. The scalar quantity given by

$$\frac{\max_i \left(x_i^{(0)} v_i^{(0)}\right)}{\min_i \left(x_i^{(0)} v_i^{(0)}\right)} \tag{10.3}$$

will be called the *condition number* associated with the homotopy system (10.2).

3. Let us denote the homotopy system of equations (10.2) more compactly by $\mathbf{H}(\mathbf{w}, \mu) = \mathbf{0}$. Its Jacobian matrix, namely, the matrix of partial derivatives of $\mathbf{H}(\mathbf{w}, \mu)$ with respect to $\mathbf{w}$, is

$$\mathbf{H}'_w = \begin{bmatrix} \mathbf{A} & \mathbf{0} & \mathbf{0} \\ \mathbf{0} & \mathbf{A}^T & \mathbf{I} \\ \mathbf{V} & \mathbf{0} & \mathbf{X} \end{bmatrix}. \tag{10.4}$$

When $\mathbf{A}$ is of full rank and the matrices $\mathbf{X}$ and $\mathbf{V}$ have positive diagonal elements, then $\mathbf{H}'_w$ is *nonsingular.*

Now, intuitively, it is plausible that the points $\mathbf{w}(\mu) = (\mathbf{x}(\mu), \pi(\mu), \mathbf{v}(\mu))$, $1 \geq \mu \geq 0$, form a *unique differentiable path* of solutions of the homotopy system (10.2), leading from the given point $\mathbf{w}^{(0)} = \left(\mathbf{x}^{(0)}, \pi^{(0)}, \mathbf{v}^{(0)}\right)$ to a solution of the original system. Moreover, the assumption that $\mathbf{w}^{(0)}$ is in the positive orthant with respect to the variables $\mathbf{x}$ and $\mathbf{v}$ implies that $\mu\mathbf{X}_0\mathbf{V}_0\mathbf{e} > \mathbf{0}, \forall\mu > 0$, and then the form of the third set of equations in (10.2) suggests that the *entire* path $\mathbf{w}(\mu)$, $\mu \in (0, 1]$, remains in the positive orthant with respect to $\mathbf{x}$ and $\mathbf{v}$. (The exception is the terminal point corresponding to $\mu = 0$, which can lie on the boundary.) To be more specific, suppose that a component of the variables $\mathbf{x}$ associated with a point on the path, say the ith, crosses its axis and thus takes on the value zero. Then the ith component of the variables $\mathbf{v}$ must assume an infinite value. *But this cannot happen if the path is shown to be bounded.* A similar argument applies with the roles of $\mathbf{x}$ and $\mathbf{v}$ reversed. The third set of equations of the square system (10.2) defining the path thus serves as a "proxy" for the nonnegativity bounds on $\mathbf{x}$ and $\mathbf{v}$ in the optimality conditions (10.1).

Path existence can be formalized following a line of verification that is standard in the theory of homotopy-based methods, which is based on the implicit function theorem as outlined in Section 3.2, and the subsidiary proof of boundedness can be obtained by reduction to a theorem due to Dikin [1974]. See Nazareth [1991, Section 2.2] for an example of this type of argument when the starting vectors are feasible; i.e., $\mathbf{q}_{x^{(0)}} = \mathbf{0}$ and $\mathbf{q}_{v^{(0)}} = \mathbf{0}$. The extension to the more general case can be achieved along similar lines. Such proofs are facilitated by log-barrier transformations of the original linear programs and their relationship to (10.2), which are discussed in the next chapter, Section 11.1. But detailed demonstrations of path existence and smoothness will not be pursued here.

In addition to definitions itemized above, we will use the following definition concerning paths:

4. The scalar $\mathbf{x}(\mu)^T\mathbf{v}(\mu)$ will be called the *"duality gap"* associated with the point $(\mathbf{w}(\mu), \mu)$ on the homotopy path. It follows from (10.2) that

$$\mathbf{x}(\mu)^T\mathbf{v}(\mu) = (\mathbf{c} + \mu\mathbf{q}_{v^{(0)}})^T\mathbf{x}(\mu) - (\mathbf{b} + \mu\mathbf{q}_{x^{(0)}})^T\pi(\mu).$$

If the components of $\mathbf{w}(\mu)$ are feasible for (P) and (D); i.e., if $\mathbf{q}_{x^{(0)}} = \mathbf{0}$ and $\mathbf{q}_{v^{(0)}} = \mathbf{0}$, then $\mathbf{x}(\mu)^T\mathbf{v}(\mu)$ is a true duality gap for the linear programs (P) and (D). Otherwise, it represents a duality gap for a linear program with suitably modified objective and right-hand side vectors. Also, by summing the third set of equations in (10.2), it follows that

$$\mu = \frac{\mathbf{x}(\mu)^T\mathbf{v}(\mu)}{\left(\mathbf{x}^{(0)}\right)^T\mathbf{v}^{(0)}}. \tag{10.5}$$

This corresponds to the choice $\mathbf{z} = \begin{bmatrix}\mathbf{0}\\ \mathbf{e}\end{bmatrix} \in R^{m+2n}$ in expression (3.12), where $\mathbf{0} \in R^{m+n}$ and $\mathbf{e} \in R^n$ is a vector of 1's.

When the LP matrix $\mathbf{A}$ is of full rank and $\mathbf{w}(\mu)$, $\mu \neq 0$, lies within the positive orthant R^n_{++} with respect to $\mathbf{x}$ and $\mathbf{v}$ then, as noted under item 3 above, the Jacobian matrix (10.4) is *nonsingular*. This, in turn, implies that *the homotopy path has no turning points*, which greatly facilitates path-following strategies to be discussed shortly, in Section 10.2. Each point $(\mathbf{w}(\mu), \mu)$ on the homotopy path is defined by a unique μ, which, from (10.5), is the ratio of the duality gap for that point and the duality gap for the initial point. Thus, $\mu \downarrow 0$ is equivalent to saying that the duality gap associated with $(\mathbf{w}(\mu), \mu)$ tends to zero.

With each point $(\mathbf{w}(\mu), \mu)$, $\mu > 0$, on the path, define a condition number

$$\max_i (x_i(\mu)v_i(\mu)) / \min_i (x_i(\mu)v_i(\mu)).$$

The third set of equations in (10.2) implies that the foregoing condition number, for any value of $\mu > 0$, is equal to the condition number (10.3) defined from the path's initial point $(\mathbf{w}(1), 1)$; i.e., the condition number associated with the homotopy system (10.2) can be obtained from any other interior point on the path.

Note that μ need not be confined to the interval $[0, 1]$. As μ is increased beyond 1, the homotopy system (10.2) is deformed away from optimal solutions of the original linear programs. Continuation of μ below zero yields solutions to (10.2) that lie outside the nonnegative $(\mathbf{x}, \mathbf{v})$ orthant. The latter noninterior case presents some interesting issues that will be pursued only very briefly in the present monograph, in Section 10.1.2.

Categorization of Interior Paths

We will refine the path terminology henceforth as follows. The path associated with the system (10.2) will be called an *interior homotopy path*. Note that it lies in a space of dimension $(m + 2n + 1)$ corresponding to the variables $(\mathbf{w}, \mu)$. Its projection into the space of variables $\mathbf{w}$ of dimension $(m + 2n)$ is called a *lifted* interior path. When the context is clear, we will often say "interior path" or "general path" or simply "path" when referring to an interior homotopy path or to its lifted counterpart.

We emphasize that points $\mathbf{w}(\mu)$ on a (lifted) interior path are *not* required to be feasible; i.e., the components of $\mathbf{w}^{(0)}$ are not required to satisfy the equality constraints of (P) and (D). For this case, it is easy to ensure that all components of the vector $\mathbf{X}_0\mathbf{V}_0\mathbf{e}$ in (10.2) are equal, which implies that the homotopy system (10.2) and associated path have the optimal condition number of unity. There are two simple ways to achieve this property, namely:

- Given any $\mathbf{x}^{(0)} > \mathbf{0}$, let $v_i^{(0)} = 1/x_i^{(0)}, \forall i$ (or vice versa), and let $\pi^{(0)} = \mathbf{0}$. We will call the associated lifted path an *optimally conditioned* (infeasible) *interior path.*
- A special case is to choose $\mathbf{x}^{(0)} = \mathbf{v}^{(0)} = \mu^{(0)}\mathbf{e}$, $\mu^{(0)} > 0$, and $\pi^{(0)} = \mathbf{0}$. The corresponding lifted path of solutions of (10.2) will then be termed a *central ray-initiated* (optimally conditioned, infeasible) *interior path.*

When the initial point $\mathbf{w}^{(0)}$ is *feasible*, the path associated with (10.2) is a *feasible interior homotopy path*, and the corresponding lifted path, which consists of feasible points, is a *feasible interior path.*

Does there exist an optimally conditioned feasible interior path? The answer is yes, and this *unique* path is called the *central path* of the linear program. A very convenient way to establish its properties of existence and uniqueness is via the log-barrier transformations of the next chapter; see Section 11.1.1 and the discussion after expression (11.4). The central path is a fundamental object of linear programming. It was introduced in an alternative setting in Section 8.4 and will be discussed further in Section 10.2. We will let the context determine whether the term "central path" applies to the homotopy path in a space of dimension $(m + 2n + 1)$ or to its lifted counterpart in the primal–dual space (or even its restriction to the space of primal or dual variables).

We illustrate the paths defined above using a simple example, similar to the one given at the beginning of Chapter 7 and depicted in Figure 7.1. (The axes are reordered here, for convenience). The corresponding dual linear program has variables $\pi \in R$ and $\mathbf{v} \in R^3$. The feasible region for the π variable is the half-line $(-\infty, \pi^*)$, where π^* is the optimal value, and the feasible region for the $\mathbf{v}$ variables is also a half-line in R^3_+. The feasible regions for both primal and dual are depicted in Figure 10.1, with the π and $\mathbf{v}$ variables shown separately. The $\mathbf{w}$-space has 7 dimensions. A (general) homotopy path is projected separately into the $\mathbf{x}$, π, and $\mathbf{v}$ spaces, and the resulting lifted paths (G) are depicted. Lifted paths are also shown for several other cases: the central path (C), a feasible interior path (F), a central ray-initiated interior path (Ia), and an optimally conditioned interior path (Ib).

For some additional detail on interior path categorization, see Nazareth [1998].

10.1.2 Noninterior Paths

Now consider an *arbitrary* initial point $\mathbf{w}^{(0)}$; i.e., its components $\mathbf{x}^{(0)}$ and $(\pi^{(0)}, \mathbf{v}^{(0)})$ are not required to have positive elements or satisfy the equality constraints of (P) or (D). Let us choose positive constants ω_i and γ_i, $i = 1, \ldots, n$, and define associated quantities

$$x_i^{+(0)} = x_i^{(0)} + \omega_i, \qquad v_i^{+(0)} = v_i^{(0)} + \gamma_i, \qquad i = 1, \ldots, n, \tag{10.6}$$

so that $\mathbf{x}^{+(0)} > \mathbf{0}$, $\mathbf{v}^{+(0)} > \mathbf{0}$, and the condition number of the diagonal matrix

$$\operatorname{diag}\left[x_1^{+(0)} v_1^{+(0)}, \ldots, x_n^{+(0)} v_n^{+(0)}\right]$$

is bounded by any desired constant, say C. Let ω and γ be n-vectors with components ω_i and γ_i.

We can now define a homotopy system associated with the starting point $\mathbf{w}^{(0)}$ as follows:

$$\begin{aligned} \mathbf{A}\mathbf{x} &= \mathbf{b} + \mu \mathbf{q}_{x^{(0)}}, \\ \mathbf{A}^T \pi + \mathbf{v} &= \mathbf{c} + \mu \mathbf{q}_{v^{(0)}}, \\ \mathbf{X}^+ \mathbf{V}^+ \mathbf{e} &= \mu \mathbf{X}_0^+ \mathbf{V}_0^+ e, \\ \mathbf{x}^+ &= \mathbf{x} + \mu \omega, \\ \mathbf{v}^+ &= \mathbf{v} + \mu \gamma, \end{aligned} \tag{10.7}$$

where $\mathbf{x}^+, \mathbf{v}^+ \in R_+^n$ and $\mu \in [0, 1]$. The components of the n-vectors $\mathbf{x}^+$ and $\mathbf{v}^+$ are newly introduced variables x_i^+ and v_i^+, respectively, and the quantities $\mathbf{X}^+$, $\mathbf{V}^+$, $\mathbf{X}_0^+$, and $\mathbf{V}_0^+$ are diagonal matrices with corresponding diagonal elements x_i^+, v_i^+, $x_i^{+(0)}$, and $v_i^{+(0)}$, $i = 1, \ldots, n$. As in (10.2), the vectors $\mathbf{q}_{x^{(0)}}$ and $\mathbf{q}_{v^{(0)}}$ are defined to be $\mathbf{q}_{x^{(0)}} = \mathbf{A}\mathbf{x}^{(0)} - \mathbf{b}$ and $\mathbf{q}_{v^{(0)}} = \mathbf{A}^T \pi^{(0)} + \mathbf{v}^{(0)} - \mathbf{c}$.

When $\mu = 1$, the homotopy system (10.7) has a solution $(\mathbf{x}^{(0)}, \pi^{(0)}, \mathbf{v}^{(0)})$ along with the vectors $\mathbf{x}^{+(0)}, \mathbf{v}^{+(0)}$ defined by (10.6). When $\mu = 0$, the system (10.7) is equivalent to the original KKT equations. For $0 \le \mu \le 1$, it defines a *noninterior homotopy path.* Using the illustrative example of the previous subsection, the associated noninterior lifted path is also depicted in Figure 10.1; see the path identified by N.

Observe that the foregoing homotopy system (10.7) can be converted to a system identical in structure to (10.2) by eliminating the original variables $\mathbf{x}$ and $\mathbf{v}$ to obtain

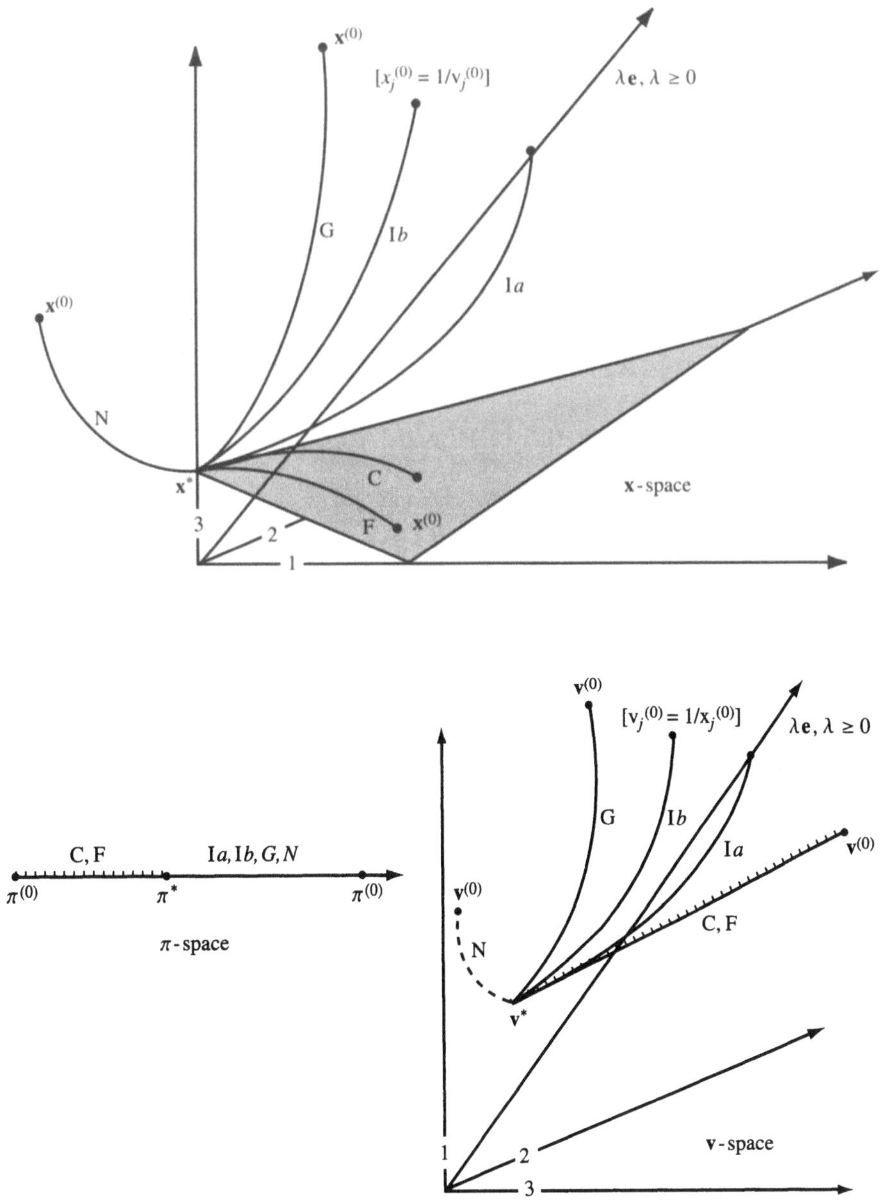

FIGURE 10.1 Illustration of paths.

$$
\begin{aligned}
\mathbf{A}\mathbf{x}^{+} &= \mathbf{b} + \mu \mathbf{q}^{+}_{x(0)}, \\
\mathbf{A}^{T}\pi + \mathbf{v}^{+} &= \mathbf{c} + \mu \mathbf{q}^{+}_{v(0)}, \\
\mathbf{X}^{+}\mathbf{V}^{+}\mathbf{e} &= \mu \mathbf{X}_0^{+}\mathbf{V}_0^{+} e,
\end{aligned}
\tag{10.8}
$$

and $\mathbf{x}^+, \mathbf{v}^+ \in R^n_+$. The vectors in the right-hand sides of the first two equations are given by $\mathbf{q}^+_{x(0)} = \mathbf{q}_{x(0)} + \mathbf{A}\omega$ and $\mathbf{q}^+_{v(0)} = \mathbf{q}_{v(0)} + \gamma$. This transformation is useful for purposes of mathematical analysis, because all previous results of Section 10.1.1 along with subsequent discussion on path-following in Section 10.2 can be carried over to the noninterior case. For conceptual and implementational purposes, however, the form (10.7) may be preferable.

We do not consider the noninterior case in any detail in this monograph. For some related references, see, for example, Chen and Harker [1993] and Burke and Xu [2000].

10.2 Path-Following Building Blocks

We now turn to the formulation of algorithms for the system (10.2) premised on the homotopy method of Sections 3.2.2–3.2.4.

From a conceptual standpoint, homotopy, or path-following,[1] techniques lie at the foundation of post-Karmarkar interior-point methods for linear programming. They have found *fundamental new expression* within the linear programming context, and these LP adaptations and developments, in turn, have considerably reilluminated usage within the traditional general *nonlinear* equations setting. The situation is analogous to that for classical log-barrier techniques of nonlinear programming, which also play a fundamental role and have found fresh expression within the LP setting, as we shall see in the next chapter.

The homotopy system (10.2) is a mixture of $(n + m)$ linear equations and n nonlinear equations that have special structure. These particular $(2n + m)$ equations, in turn, give the path-following approach some very distinctive characteristics in the LP setting. The most prominent of these characteristics is the existence and special role of the *central path*: the unique optimally conditioned lifted path within the feasible polytope of a linear program and its dual. One must be careful, however, not to overly diminish the status of other homotopy paths discussed earlier (and their lifted counterparts), because they can play a role very similar to that of the central path within a path-following algorithm, as will be seen in subsequent discussion in this chapter. Put another way, the central path is a fundamental object of linear programming, but only a *first among equals*.

As discussed in Section 10.1.1, LP homotopy interior paths have no turning points when $\mathbf{A}$ is of full rank, and they are relatively simple to characterize. Convenient expressions can be found for their tangents and curvature, which will be given in Sections 10.2.2–10.2.3. When devising path-following algorithms, one can draw on the extensive knowledge of homotopy tech-

[1] We will use these two names interchangeably in this monograph.

niques surveyed in Chapter 3, in particular, techniques of the predictor–corrector variety. The latter are capable of significant enhancement in the LP setting, because of the linearity/nonlinear structure described above. These techniques are the "fundamentally different algorithms" of the second quotation in Section 3.2.4 and the subsequent considerations of that section.

We now discuss the set of *simple building blocks* from which such path-following algorithms can be composed.

10.2.1 Notation

For direction vectors below, we use the symbols $\mathbf{p}$ for predictor, $\mathbf{s}$ for second-order corrector, $\mathbf{n}$ for Newton corrector, and $\mathbf{c}$ for centering component.

Points or iterates of an algorithm that lie *off* the global homotopy path that is being followed are distinguished by an overbar, for example, $\overline{\mathbf{w}}$ or $\overline{\mathbf{w}}^{(k)}$. Points that lie *on* the path always have a superscript or a parameter μ associated with them, and the overbar is omitted, for example, $\mathbf{w}^{(0)}$ or $\mathbf{w}(\mu)$.

An arbitrary point of the primal–dual space of variables is denoted by $\mathbf{w} = (\mathbf{x}, \pi, \mathbf{v})$. The interior of the primal and dual feasible polytopes is denoted by $\mathcal{F}^0$.

We use $\mathbf{q}$ for residuals associated with the original KKT system (10.1) and $\mathbf{r}$ for residuals associated with the homotopy system (10.2). These have three main components, corresponding to the three sets of equations in (10.1) and (10.2), and the first two components are identified by the subscripts x and v, respectively (with an attached overbar when defined at points off the path).

10.2.2 Euler Predictor

Differentiating the homotopy system (10.2), namely $\mathbf{H}(\mathbf{w}(\mu), \mu) = \mathbf{0}$, with respect to μ yields the homotopy differential equation (HDE) analogous to (3.11). Linearizing the HDE at the initial point $\mathbf{w}^{(0)}$ yields the equations that define the tangent vector to the lifted path along the direction of increasing μ. The Euler predictor direction is the negative of this vector. Its components $\mathbf{p}_{w^{(0)}} = (\mathbf{p}_{x^{(0)}}, \mathbf{p}_{\pi^{(0)}}, \mathbf{p}_{v^{(0)}})$ are along the (negative) tangent vector to the lifted path at the point $\mathbf{w}^{(0)} = (\mathbf{x}^{(0)}, \pi^{(0)}, \mathbf{v}^{(0)})$ and are defined by the following set of linear equations:

$$\begin{aligned} \mathbf{A}\mathbf{p}_{x^{(0)}} &= -\mathbf{q}_{x^{(0)}}, \\ \mathbf{A}^T\mathbf{p}_{\pi^{(0)}} + \mathbf{p}_{v^{(0)}} &= -\mathbf{q}_{v^{(0)}}, \\ \mathbf{X}_0\mathbf{p}_{v^{(0)}} + \mathbf{V}_0\mathbf{p}_{x^{(0)}} &= -\mathbf{X}_0\mathbf{V}_0\mathbf{e}, \end{aligned} \tag{10.9}$$

where in the right-hand side of the last equation, the components of the vectors are multiplied pairwise to form a vector. The vectors and diagonal matrices above were defined in Section 10.1.1. Note that the matrix in the equations defining the Euler predictor is the Jacobian matrix (10.4) at the point $\mathbf{w} = \mathbf{w}^{(0)}$.

Consider any other point *on the path* leading from $(\mathbf{w}^{(0)}, 1)$ to $(\mathbf{w}^*, 0)$, say the point $(\mathbf{w}(\mu), \mu)$, $0 < \mu < 1$. Recalling the discussion on restart strategies of Section 3.2.4, suppose one *restarted* the homotopy at this point. By uniqueness of the original path, this must correspond simply to a reparameterization: rescaling the initial value of the parameter to 1. The restarted lifted path must overlap the original path to $\mathbf{w}^*$. Thus the (negative) tangent vector $(\mathbf{p}_x, \mathbf{p}_\pi, \mathbf{p}_v)$ at any point $(\mathbf{x}(\mu), \pi(\mu), \mathbf{v}(\mu))$ of the lifted path corresponding to the original system (10.2) is given by (10.9) with a simple transcription of symbols. For later reference, let us restate these equations:

$$\begin{aligned} \mathbf{A}\mathbf{p}_x &= -\mathbf{q}_x, \\ \mathbf{A}^T\mathbf{p}_\pi + \mathbf{p}_v &= -\mathbf{q}_v, \\ \mathbf{X}\mathbf{p}_v + \mathbf{V}\mathbf{p}_x &= -\mathbf{X}\mathbf{V}\mathbf{e}, \end{aligned} \tag{10.10}$$

where the diagonal matrices $\mathbf{X}$ and $\mathbf{V}$ have diagonal elements given by the components of $\mathbf{x}(\mu)$ and $\mathbf{v}(\mu)$, and the right-hand-side vectors are defined by $\mathbf{q}_x = \mathbf{A}\mathbf{x}(\mu) - \mathbf{b}$ and $\mathbf{q}_v = \mathbf{A}^T\pi(\mu) + \mathbf{v}(\mu) - \mathbf{c}$.

The following properties of the Euler predictor are easy to verify, and details are left to the reader:

- The Euler predictor direction at the point $\mathbf{w}(\mu)$ is parallel to the search direction defined by Newton's method applied to the KKT system of optimality equations (10.1) at the point $\mathbf{w}(\mu)$. In particular, this equivalence of directions holds at the initiating point $\mathbf{w}^{(0)}$. Note that the only restriction placed on $\mathbf{w}^{(0)}$ is that its components satisfy $\mathbf{x}^{(0)} > \mathbf{0}$ and $\mathbf{v}^{(0)} > \mathbf{0}$.
- Suppose that the path under consideration is the *central* path. Then the Euler predictor at any point on this path lies along the (negative) tangent vector to the path, and its primal and dual components are parallel to the primal affine-scaling and the dual affine-scaling directions, respectively, as defined in Sections 8.1 and 8.2. These directions coincide with the corresponding primal and dual components of the primal–dual affine-scaling directions given in Section 8.3.
- In the case of a *feasible* interior path, the Euler predictor direction $\mathbf{p}_w$ at a point $\mathbf{w}(\mu)$ on the path is the primal–dual affine-scaling direction as developed in expressions (8.18) and (8.19) with the appropriate transcription of symbols; i.e.,

$$\mathbf{X} \leftrightarrow \mathbf{D}_X; \qquad \mathbf{V} \leftrightarrow \mathbf{D}_V. \tag{10.11}$$

But note that the primal and dual components of $\mathbf{p}_w$ are now *distinct* from the primal affine-scaling and dual affine-scaling directions defined at the feasible interior primal and dual components of $\mathbf{w}(\mu)$.

- In the case of a *general* interior path; i.e., a path that need not lie within the feasible polytope of (P) and (D), the Euler predictor direction at $\mathbf{w}(\mu)$ is defined in terms of a primal–dual affine-scaling diagonal *matrix*, but it does *not* correspond to the direction that is defined in expressions (8.18) and (8.19), again with the transcription of symbols (10.11). It is standard terminology to continue to use the name primal–dual affine-scaling direction for the Euler predictor corresponding to the general (infeasible) interior case.

10.2.3 Second-Order Corrector

The vector $\mathbf{s}_w = (\mathbf{s}_x, \mathbf{s}_\pi, \mathbf{s}_v)$ defining the second derivative to the lifted path at the point $(\mathbf{x}(\mu), \pi(\mu), \mathbf{v}(\mu))$ can be obtained along lines analogous to the derivation of the first derivative above. It is given by solving the following set of linear equations:

$$\begin{aligned} \mathbf{A}\mathbf{s}_x &= \mathbf{0}, \\ \mathbf{A}^T\mathbf{s}_\pi + \mathbf{s}_v &= \mathbf{0}, \\ \mathbf{X}\mathbf{s}_v + \mathbf{V}\mathbf{s}_x &= -2\mathbf{p}_x\mathbf{p}_v, \end{aligned} \tag{10.12}$$

where in the right-hand side of the last equation, the components of the Euler predictor vector are multiplied pairwise to form a new vector. The diagonal matrices $\mathbf{X}$ and $\mathbf{V}$ are defined as in the Euler predictor equations (10.10).

10.2.4 Newton Corrector

Consider a point, denoted by $\overline{\mathbf{w}} = (\overline{\mathbf{x}}, \overline{\pi}, \overline{\mathbf{v}})$, that lies *off* the (general) homotopy path and an associated value μ. The Newton corrector $(\mathbf{n}_{\overline{x}}, \mathbf{n}_{\overline{\pi}}, \mathbf{n}_{\overline{v}})$ at this point is obtained, analogously to its definition in expression (3.2), by linearizing the system (10.2). It is given by the following system of linear equations:

$$\begin{aligned} \mathbf{A}\mathbf{n}_{\overline{x}} &= -\mathbf{r}_{\overline{x}}, \\ \mathbf{A}^T\mathbf{n}_{\overline{\pi}} + \mathbf{n}_{\overline{v}} &= -\mathbf{r}_{\overline{v}}, \\ \overline{\mathbf{V}}\mathbf{n}_{\overline{x}} + \overline{\mathbf{X}}\mathbf{n}_{\overline{v}} &= -\overline{\mathbf{X}}\overline{\mathbf{V}}\mathbf{e} + \mu\mathbf{X}_0\mathbf{V}_0\mathbf{e}, \end{aligned} \tag{10.13}$$

where $\mathbf{e} \in R^n$ denotes the vector of all ones, and $\overline{\mathbf{X}}$ and $\overline{\mathbf{V}}$ denote diagonal matrices with positive diagonal elements defined by the components of $\overline{\mathbf{x}}$ and $\overline{\mathbf{v}}$, respectively. The right-hand sides of the first two equations are residual vectors, namely,

$$\mathbf{r}_{\overline{x}} = \mathbf{A}\overline{\mathbf{x}} - \mathbf{b} - \mu\mathbf{q}_{x^{(0)}} = \mathbf{q}_{\overline{x}} - \mu\mathbf{q}_{x^{(0)}},$$

where $\mathbf{q}_{\overline{x}} = \mathbf{A}\overline{\mathbf{x}} - \mathbf{b}$ and

$$\mathbf{r}_{\overline{v}} = \mathbf{A}^T\overline{\pi} + \overline{\mathbf{v}} - \mathbf{c} - \mu\mathbf{q}_{v^{(0)}} = \mathbf{q}_{\overline{v}} - \mu\mathbf{q}_{v^{(0)}}$$

with $\mathbf{q}_{\overline{v}} = \mathbf{A}^T\overline{\pi} + \overline{\mathbf{v}} - \mathbf{c}$.

Denote the Newton corrector direction more compactly by the vector $\mathbf{n}_{\overline{w}} = (\mathbf{n}_{\overline{x}}, \mathbf{n}_{\overline{\pi}}, \mathbf{n}_{\overline{v}})$. If the point $\overline{\mathbf{w}}$ is infeasible, but a *feasible* homotopy path is being followed, then the residuals in the right-hand side of the Newton corrector equations (10.13) correspond to the residuals for $\overline{\mathbf{w}}$ with respect to the original KKT equations (10.1). And in this case, if a *full*, i.e., a unit, step is taken along the Newton corrector direction, then the resulting point $\overline{\mathbf{w}} + \mathbf{n}_{\overline{w}}$ becomes *feasible.*

10.2.5 Restarts

Restarting and reparameterization of a homotopy have been described in Section 3.2.4, and they can be applied to the homotopy system (10.2). In particular, restarting allows two different HDEs to be used simultanously within a bipath-following strategy. This holds the key to formulating effective algorithms in the linear programming setting, as we shall see in the next section.

Consider a restart of the homotopy defined by (3.14), applied to the homotopy system (10.2) at the point $\overline{\mathbf{w}} = (\overline{\mathbf{x}}, \overline{\pi}, \overline{\mathbf{v}})$. For convenience of discussion, let us identify the original homotopy system and associated HDE by $\mathcal{A}$ and the restarted homotopy system and its HDE by $\mathcal{B}$. The Euler predictor (EP) for homotopy $\mathcal{B}$ at the point $\overline{\mathbf{w}}$ is given by (10.9) with the components of the vector $\mathbf{w}^{(0)} = \left(\mathbf{x}^{(0)}, \pi^{(0)}, \mathbf{v}^{(0)}\right)$ replaced by the components of $\overline{\mathbf{w}} = (\overline{\mathbf{x}}, \overline{\pi}, \overline{\mathbf{v}})$, and the diagonal matrices $\mathbf{X}_0$ and $\mathbf{V}_0$ replaced by $\overline{\mathbf{X}}$ and $\overline{\mathbf{V}}$, respectively. The Newton corrector (NC) for homotopy $\mathcal{A}$ is given by (10.13).

It is useful to define the centering component (CC) $\mathbf{c}_{\overline{w}} = (\mathbf{c}_{\overline{x}}, \mathbf{c}_{\overline{\pi}}, \mathbf{c}_{\overline{v}})$ for the original homotopy $\mathcal{A}$ at the point $\overline{\mathbf{w}} = (\overline{\mathbf{x}}, \overline{\pi}, \overline{\mathbf{v}})$ as the solution of the following system of linear equations:

$$\begin{aligned} \mathbf{A}\mathbf{c}_{\overline{x}} &= \mathbf{q}_{x^{(0)}}, \\ \mathbf{A}^T\mathbf{c}_{\overline{\pi}} + \mathbf{c}_{\overline{v}} &= \mathbf{q}_{v^{(0)}}, \\ \overline{\mathbf{V}}\mathbf{c}_{\overline{x}} + \overline{\mathbf{X}}\mathbf{c}_{\overline{v}} &= \mathbf{X}_0\mathbf{V}_0\mathbf{e}. \end{aligned} \tag{10.14}$$

Note that this system differs from the system defining the Newton corrector only in its right-hand-side vector. Also, the Euler predictor for homotopy $\mathcal{B}$ at $\overline{\mathbf{w}}$ has the same left-hand-side matrix as the Newton corrector and the centering component for homotopy $\mathcal{A}$. It then follows immediately that

$$[\text{NC for } \mathcal{A} \text{ at } \overline{\mathbf{w}}] = [\text{EP for } \mathcal{B} \text{ at } \overline{\mathbf{w}}] + \mu[\text{CC for } \mathcal{A} \text{ at } \overline{\mathbf{w}}]. \tag{10.15}$$

The relationship (10.15) will turn out to be very useful in the formulation of algorithms later. An important special case of (10.15) arises when the homotopy paths are *feasible*. In this case all residuals in the equations defining the EP, NC, and CC are zero. If, additionally, the homotopy path $\mathcal{A}$ is the central path, then the diagonal matrix $\mathbf{X}_0\mathbf{V}_0$ is replaced by the identity matrix.

It is also useful to note when a bipath strategy is employed that the option exists of using a starting point for an algorithm that does *not lie on the homotopy path* $\mathcal{A}$ being followed. This corresponds to performing a restart and defining a homotopy system $\mathcal{B}$ at the very outset. For example, homotopy $\mathcal{A}$ could correspond to the central path, and an algorithm could be (re-) started at a point that lies within some prespecified neighborhood of the central path.

10.3 Numerical Illustration

A pure Euler predictor strategy treats the HDE (3.11) as an initial value problem along lines discussed in Chapter 3, where $\mathbf{H}$, in the present context, is defined by (10.2) and the Euler predictor by (10.9). At a point that lies off the path, the Euler predictor direction would be derived from the same HDE and would be tangential to a path arising from use of *a different initial condition.* Small steps would be taken within a "tubular" neighborhood of the original homotopy path in order to follow the latter accurately.

A second basic strategy is to take an Euler predictor step along the direction defined by (10.9) and then immediately to restart the homotopy at the new iterate. Now the path and associated HDE are changed at each iteration, and the "terminal value" aspect of path following, as discussed in Section 3.2.4, is addressed. We have noted that this iteration is equivalent to Newton's method applied to the original KKT conditions of the linear program, and the convergence of the pure Newton iteration is not guaranteed, as is well known.

The algorithm implemented in the numerical illustration of this section extends the foregoing terminal-value strategy by inserting a Newton corrector phase between the Euler predictor phase and the restart as follows:

Algorithm EP/NC/R: Given an initial point $\mathbf{w}^{(0)} = \left(\mathbf{x}^{(0)}, \pi^{(0)}, \mathbf{v}^{(0)}\right)$ that need not be feasible for (P) and (D) but must satisfy $\mathbf{x}^{(0)} > \mathbf{0}$ and $\mathbf{v}^{(0)} > \mathbf{0}$:

1. *Predictor Phase*: Compute the Euler predictor $\mathbf{p}_{w^{(0)}} = (\mathbf{p}_{x^{(0)}}, \mathbf{p}_{\pi^{(0)}}, \mathbf{p}_{v^{(0)}})$ from (10.9). Take a step t_P, $0 < t_P \leq 1$, along it to a new point, $\overline{\mathbf{w}} = (\overline{\mathbf{x}}, \overline{\pi}, \overline{\mathbf{v}})$; i.e., $\overline{\mathbf{w}} = \mathbf{w}^{(0)} + t_P\mathbf{p}_{w^{(0)}}$. The step length t_P is obtained by computing the largest steps t_x and t_v in the interval

[0, 1] such that $\overline{\mathbf{x}} \geq \mathbf{0}$ and $\overline{\mathbf{v}} \geq \mathbf{0}$ and defining $t_P = \tau * \min[t_x, t_v]$ with $\tau \in (0, 1)$. Correspondingly, compute the parameter $\mu = (1 - t_P)$.

2. *Corrector Phase*: The Euler predictor direction is obtained from a linearization of the system (10.2). The first two equations of the latter system are already linear, and thus the two equations will be satisfied *exactly* by the point $\overline{\mathbf{w}}$ with the homotopy parameter at the value of μ computed above. In other words, the relations $\mathbf{A}\mathbf{x}^{(0)} = \mathbf{b} + \mathbf{q}_{x^{(0)}}$ and $\mathbf{A}\mathbf{p}_{x^{(0)}} = -\mathbf{q}_{x^{(0)}}$ imply that

$$\mathbf{A}\overline{\mathbf{x}} = \mathbf{b} + (1 - t_P)\mathbf{q}_{x^{(0)}} = \mathbf{b} + \mu\mathbf{q}_{x^{(0)}}. \tag{10.16}$$

Similarly,

$$\mathbf{A}^T\overline{\pi} + \overline{\mathbf{v}} = \mathbf{c} + (1 - t_P)\mathbf{q}_{v^{(0)}} = \mathbf{c} + \mu\mathbf{q}_{v^{(0)}}. \tag{10.17}$$

The equations (10.13) defining the Newton corrector direction $\mathbf{n}_{\overline{w}}$ must therefore have zero right-hand-side residual vectors, and a Newton corrector step along the direction computed from (10.13) will produce an iterate that continues to satisy the first two sets of equations. Take a common step t_C no greater than unity that does not violate a bound constraint along the direction, in an analogous way to the predictor phase above. Update the current iterate and denote it again by $\overline{\mathbf{w}}$. Then repeat, if necessary. Since the first two sets of linear equations in (10.2) are already satisfied, the sum of squares of the residuals for the third set of equations of the homotopy system, namely, the function $F_\mu(\overline{\mathbf{w}}) = \frac{1}{2}r_\mu(\overline{\mathbf{w}})^T r_\mu(\overline{\mathbf{w}})$ with $r_\mu(\overline{\mathbf{w}}) = (\overline{\mathbf{X}\mathbf{V}} - \mu\mathbf{X}_0\mathbf{V}_0)\mathbf{e}$, is a natural *merit function* for the Newton iteration. One or more Newton corrector steps are taken in sequence, and the corrector phase is terminated when the merit function is sufficiently small.

3. *Restart Phase*: Perform a restart by setting $\mathbf{w}^{(0)}$ to the current iterate $\overline{\mathbf{w}}$ at the end of the corrector phase. This is the most aggressive choice of periodic restart strategy, because the path and its associated HDE are changed after every predictor–corrector cycle.

In our implementation of the foregoing Algorithm EP/NC/R, the linear program being solved is assumed to be sparse and specified in standard MPS format—see, for example, Nazareth [1987]—but only nonnegative lower bounds on variables are allowed (instead of the more general lower and upper bounds of standard MPS format). The linear program is input and represented as a column list/row index packed data structure using subroutines described in Nazareth [1986f], [1987]. All operations involving the linear programming matrix $\mathbf{A}$ in (P) or (D) in (8.1) employ this packed data structure. The linear equations defining the Euler predictor and the Newton corrector directions are solved by a simple linear CG algorithm without preconditioning—see Section 5.1—and the procedure is

TABLE 10.1. SHARE2B: 97 rows, 79 columns (excluding slacks), 730 nonzero elements, $z^* = -415.732$.

It.	$\mathbf{c}^T\mathbf{x}^{(0)}$	$\mathbf{b}^T\pi^{(0)}$	$\Vert\mathbf{q}_{x^{(0)}}\Vert/\Vert\mathbf{b}\Vert$	$\Vert\mathbf{q}_{v^{(0)}}\Vert/\Vert\mathbf{c}\Vert$	$\max\left[x_i^{(0)}v_i^{(0)}\right]$	n_C
2	−1762.69	−22735.4	6698.03	110.590	3370.96	3
3	−897.465	−21380.3	2257.86	37.2790	1104.40	3
4	−590.548	−15098.9	739.723	12.2134	416.605	3
5	−439.873	−8861.92	279.040	4.60717	138.440	3
6	−373.861	−3784.46	92.7267	1.53099	36.0400	3
7	−358.176	−1131.10	24.1394	0.398560	6.08006	4
8	−377.788	−536.579	4.07240	0.067328	1.13203	4
9	−394.126	−464.520	0.758229	0.012519	0.491282	3
10	−405.324	−436.222	0.329061	0.005433	0.212608	3
11	−412.181	−422.425	0.142403	0.002351	0.071165	3
12	−414.612	−417.649	0.047666	0.000787	0.021150	3
13	−415.324	−416.351	0.014166	0.000234	0.007168	3
14	−415.568	−415.961	0.004801	0.000079	0.002747	2
15	−415.679	−415.803	0.001840	0.000030	0.000860	2
16	−415.731	−415.743	0.000575	0.000010	0.000134	2
17	−415.733	−415.733	0.002324	0.000002	0.000009	2
18	−415.731	−415.732	0.000272	0.000000	0.000000	1
19	−415.732	−415.732	0.000015	0.000000	0.000000	1
20	−415.732	−415.732	0.000008	0.000000	0.000000	1
21	−415.732	−415.732	0.000010	0.000000	0.000000	–

terminated when the norm of the residual of the linear system, relative to the norm of its initial right-hand side, is no greater than 10^{-6}. A standard choice for the parameter τ in the predictor phase is a number in the range $[0.95, 0.99]$, and in these experiments, $\tau = 0.95$ is used. The Newton iteration in the corrector phase is terminated when the norm of the merit function, relative to $\max_{i=1,\ldots,n} \overline{x}_i\overline{v}_i$, falls below 10^{-4}. For further details on the implementation, in particular, the choice of starting point and the step-length computation, see Nazareth [1996b].

Results for a realistic problem from the Netlib collection (Gay [1985]) are shown in Table 10.1. Problem statistics for this problem, called SHARE2B, are stated in the caption of the table. Its optimal value is z^*. At the end of each predictor, corrector(s), and restart cycle of the above Algorithm EP/NC/R, the table shows the attained objective values[2] for the primal

[2]The small loss of monotonicity in row 17 of the fourth column of the table is the result of insufficient accuracy in the computation of the predictor direction. Also, in this example, the maximum and minimum values of the quantities $x_i^{(0)}v_i^{(0)}$ are almost identical throughout, and the central-ray-initiated path is followed closely. In other examples reported in Nazareth [1996b], these quantities differ significantly.

and dual linear programs, the relative infeasibility for primal and dual, the largest element in the "duality gap," and the number of Newton corrector steps taken (n_C). These results are representative of performance on several other test problems from the Netlib collection, for which similar tables are given in Nazareth [1996b]. In addition, many refinements of the implementation are possible, for example, use of different steps in primal and dual space, extrapolation, inexact computation of search vectors, and so on. Details can be found in Nazareth [1996b].

It is evident that convergence can be ensured in a *conceptual* version of Algorithm EP/NC/R, where $\tau > 0$ can be chosen to be as small a fraction as is desired for theoretical purposes in the predictor phase, and the homotopy system (10.2) is solved *exactly* in the corrector phase. (The restart then corresponds to a reparameterization of the original homotopy.) In an *implementable* version of the algorithm, where larger values of τ are used and relatively high accuracy is requested in the residual of the corrector phase, the homotopy path will *not* change substantially between consecutive cycles. Although not formalized, a proof of convergence can be derived. This is the version of Algorithm EP/NC/R that is implemented in the foregoing numerical illustration.

From a practical standpoint, it is common to use values of τ very close to unity, for example, 0.99, and to enforce the condition $n_C = 1$. In this case the path can change drastically after a restart. Convergence requires the introduction of a bipath-following strategy so that the ability to correct to a guiding homotopy path from the original starting point is retained. We consider strategies of this type in the next subsection.

10.4 Path-Following Algorithms

10.4.1 Notation

We now use the more general notation of Chapter 3 for defining iterates of an algorithm; i.e., iteration numbers will be specified explicitly.

Points generated by a conceptual algorithm that lie *on* the homotopy path being followed will be denoted by $\mathbf{w}^{(k)} = \left(\mathbf{x}^{(k)}, \pi^{(k)}, \mathbf{v}^{(k)}\right)$. A homotopy parameter value will often be specified explicitly, for example, $\mathbf{x}\left(\mu^{(k)}\right)$, highlighting the fact that the point is a *function* of the homotopy parameter μ.

Points that lie *off* the path will be denoted by $\overline{\mathbf{w}}^{(k)} = \left(\overline{\mathbf{x}}^{(k)}, \overline{\pi}^{(k)}, \overline{\mathbf{v}}^{(k)}\right)$. If a target homotopy parameter value, say $\overline{\mu}^{(k)}$, is explicitly *associated* with the point, then, as in Chapter 3, it could be denoted by $\overline{\mathbf{x}}\left[\overline{\mu}^{(k)}\right]$. Note that the point is *not* a function of the parameter. However, this extension of the notation will rarely be needed here.

Within the sets of linear equations defining the Euler predictor, Newton corrector, and other quantities in Section 10.2, iteration numbers will also be attached; for example, $\mathbf{X}$ will be replaced by $\mathbf{X}_k$, and $\mathbf{V}$ by $\mathbf{V}_k$. *Henceforth, in this chapter, the symbols* $\mathbf{X} = \mathrm{diag}[x_1, \ldots, x_n]$ *and* $\mathbf{V} = \mathrm{diag}[v_1, \ldots, v_n]$ *will be used to denote diagonal matrices at arbitrary points* $\mathbf{x} = (x_1, \ldots, x_n)$ *and* $\mathbf{v} = (v_1, \ldots, v_n)$, *respectively.*

10.4.2 Strategies

Figure 10.2 gives a sample of the many different path-following strategies that can be composed from the building blocks described in Section 10.2. *Lifted* paths are depicted. The strategies all employ the same basic template, and the continuous lines correspond to the paths used in a particular strategy. (The dotted lines in the depiction of that strategy can be ignored.) The initiating point for a path is denoted by $\left(\mathbf{x}^{(0)}, \pi^{(0)}, \mathbf{v}^{(0)}\right)$.

Paths can be feasible or infeasible, and a variety of such paths for a simple linear programming example, similar to the one of Figure 7.1, were illustrated earlier in Figure 10.1. Now Figure 10.2 gives a more generic depiction for six different path-following strategies that are discussed in the remainder of this section, and again in Section 10.4.4.

Strategy 1 and *Strategy 2* are embedding, unipath-following strategies that use a direct vertical-predictor decrease in the parameter, followed by one or more Newton corrector steps initiated from the current iterate $\overline{\mathbf{w}}^{(k)}$. *Strategy 1* is guided by the central path, and *Strategy 2* is guided by the path from the initial point $\mathbf{w}^{(0)}$.

Strategy 3 corresponds to the standard form of predictor–corrector unipath-following, where the Euler predictor is the (negative) tangent to a path through the current iterate that is determined from the *original* homotopy differential equation associated with the starting point $\mathbf{w}^{(0)}$. The latter point also determines the path that guides one or more Newton corrector steps, initiated at the point obtained from the predictor step. In *Strategy 4*, the path emanating from the current iterate defines both the Euler predictor and guides the Newton corrector steps. This is the unipath strategy used in the numerical illustration of Section 10.3.

Strategy 5 and *Strategy 6* are bipath-following strategies. In *Strategy 5*, the Euler predictor is the (negative) tangent to a path emanating from the current iterate $\overline{\mathbf{w}}^{(k)}$, and the central path guides the Newton corrector steps. In *Strategy 6*, the Euler predictor is the same as the one used in *Strategy 5*, but the Newton corrector steps are guided by the homotopy path from the initial point.

Numerous variants arise depending on whether or not the initial point is feasible[3] for the primal and dual problems, whether the initial point is additionally chosen to define an optimally conditioned or a central-ray-initiated path, and whether the Newton corrector is restricted to a *single*

---

[3]The primal variables and dual slacks must, of course, have positive components.

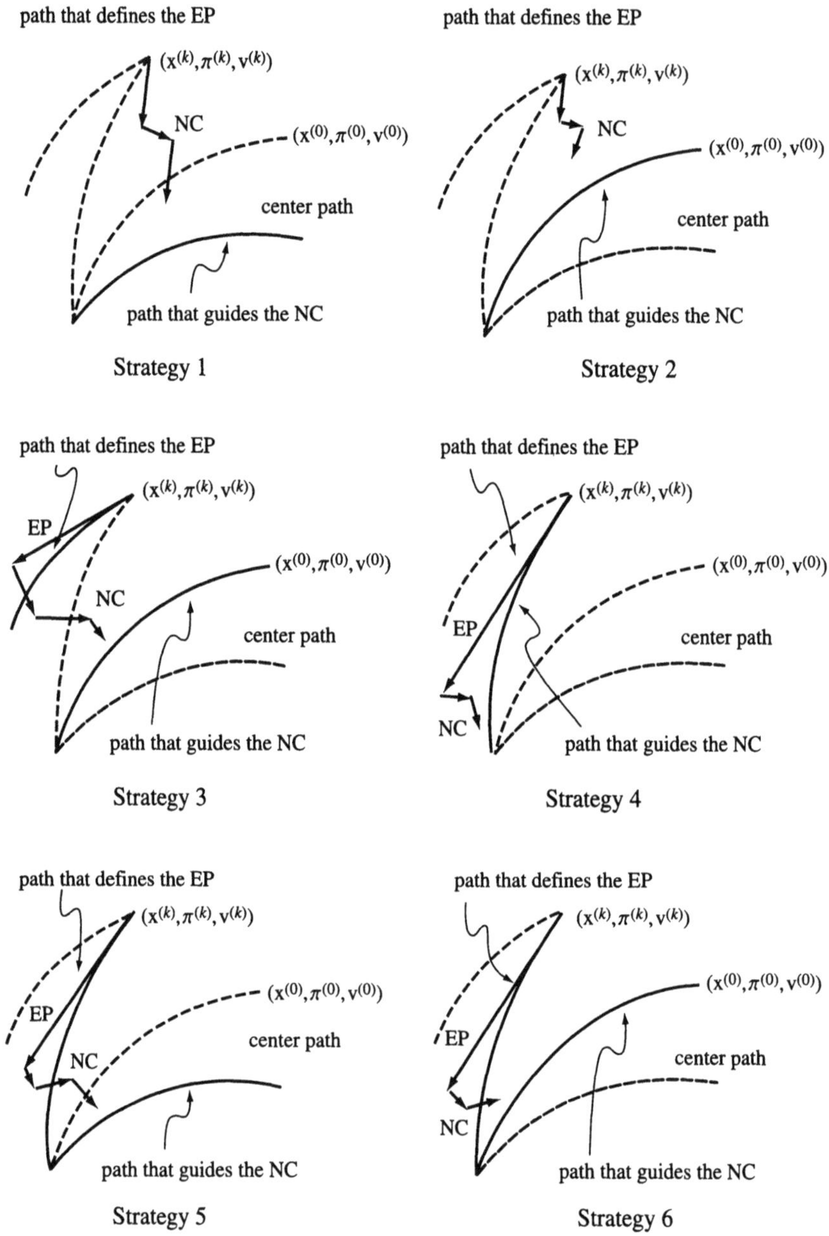

FIGURE 10.2 Strategies.

step. In all cases, an important issue in defining an algorithm is the choice of step lengths along predictor and corrector directions.

The foregoing discussion by no means exhausts the rich variety of path-following strategies that have been proposed and analyzed in the literature, but *it illustrates how they can be easily apprehended* within the path-following framework considered in this section. It is interesting to note that the most basic form of homotopy path-following, corresponding to *Strategy 3* of Figure 10.2 and patterned on standard homotopy path-following theory, is only one of many used nowadays in the linear programming setting.

The homotopy or path-following approach has been promulgated from the outset in the sequence of articles of Nazareth [1986d], [1991], [1995a], [1996b] and Sonnevend et al. [1986], [1990], [1991]. The connection between Karmarkar's interior-point approach and homotopy (path-following) techniques was first noted in Nazareth [1986d]. This connection and an important relationship between interior-point techniques and log-barrier transformations of a linear program (discussed in the next chapter) served to motivate the *landmark contribution* of Megiddo [1989] on path-following in a primal–dual setting. This article, in turn, set the stage for a variety of primal–dual path-following algorithms by many different researchers.

Although not pursued in this monograph, strategies for *noninterior* path-following as described in Section 10.1.2 can be formulated along lines similar to the strategies described above, and they deserve further investigation.

10.4.3 Measures and Targets

The formulation of an algorithm based on the foregoing strategies requires measures of the "worth" of an iterate relative to a path being followed. This is the topic of the present subsection.

Merit Functions

In the Newton corrector phase, the Newton iteration defined by (3.2) is applied to the homotopy system (10.2) for given $\mu = \mu^{(k)} > 0$, starting from the point $\overline{\mathbf{w}}^{(k)}$. The Newton iteration has an associated merit function like (3.3) for which its search direction is a direction of descent. Thus the Newton corrector direction (10.13) is a direction of descent for the sum of squares of residuals merit function, say $F_\mu(\mathbf{w}) = \frac{1}{2}\mathbf{H}(\mathbf{w}, \mu^{(k)})^T\mathbf{H}(\mathbf{w}, \mu^{(k)})$, where $\mathbf{H}$ is defined prior to expression (10.4). The Jacobian matrix of the system is nonsingular at any interior point, and the gradient vector mapping ∇F_μ can vanish only at a minimizing point of F_μ. Thus F_μ is a natural merit function for globalizing Newton's method in a corrector phase, because difficulties illustrated by the examples of Chapter 3 cannot arise.

Let us focus on the feasible case so that the first two sets of equations of (10.2) are satisfied exactly. Then the merit function involves only residuals corresponding to the third set of equations, and thus

$$F_\mu(\mathbf{w}) = \tfrac{1}{2}r_\mu(\mathbf{w})^T r_\mu(\mathbf{w}), \tag{10.18}$$

where

$$r_\mu(\mathbf{w}) = (\mathbf{XV} - \mu^{(k)}\mathbf{X}_0\mathbf{V}_0)\mathbf{e} \tag{10.19}$$

and $\mathbf{X} = \mathrm{diag}[x_1, \ldots, x_n]$, $\mathbf{V} = \mathrm{diag}[v_1, \ldots, v_n]$, and $\mathbf{X}_0$ and $\mathbf{V}_0$ are defined immediately after (10.2). The Newton corrector (10.13) will have zero residuals in the right-hand side corresponding to the first two equations, and it is a direction of descent for the merit function (10.18).

Targets on the Homotopy Path

We have also seen in expression (3.12) and its application in the present setting, namely, (10.5) and discussion immediately afterwards, that the size of the duality gap for any point *on* the path relative to the size of the starting duality gap gives the associated value of μ. This provides a natural way to *associate a target point* on a feasible homotopy path with any *feasible* iterate $\overline{\mathbf{w}}^{(k)}$ that lies off the path, namely,

$$\overline{\mu}^{(k)} = \frac{\left(\overline{\mathbf{x}}^{(k)}\right)^T \overline{\mathbf{v}}^{(k)}}{\left(\mathbf{x}^{(0)}\right)^T \mathbf{v}^{(0)}}. \tag{10.20}$$

Also,

$$\left(\overline{\mathbf{x}}^{(k)}\right)^T \overline{\mathbf{v}}^{(k)} = \mathbf{c}^T \overline{\mathbf{x}}^{(k)} - \mathbf{b}^T \overline{\pi}^{(k)}.$$

Consider a horizontal Newton corrector iteration that is initiated at the point $\left(\overline{\mathbf{w}}^{(k)}, \overline{\mu}^{(k)}\right)$ in the space of dimension $(m+2n+1)$ and restricted to a horizontal hyperplane defined by $\eta^T \mathbf{w} = \overline{\mu}^{(k)}$, where $\eta \in R^{m+2n+1}$ is a unit vector with 1 in its last position and zeros elsewhere. This corresponds to the Newton iteration applied to the system (10.2) in the space of dimension $(m + 2n)$, with $\mu = \overline{\mu}^{(k)}$ and iterates confined to lie in a hyperplane, say $\mathcal{H}\left(\overline{\mathbf{w}}^{(k)}\right)$, that passes through $\overline{\mathbf{w}}^{(k)}$ and has a normal vector $(\mathbf{c}, -\mathbf{b}, \mathbf{0})$.

A more ambitious target would be given by replacing $\overline{\mu}^{(k)}$ by $\sigma\overline{\mu}^{(k)}$, where $\sigma \in (0, 1]$. This corresponds to *shifting* the above hyperplanes in their respective spaces.

Neighborhoods

Let us confine attention to points $\mathbf{w}$ in the foregoing hyperplane $\mathcal{H}\left(\overline{\mathbf{w}}^{(k)}\right)$, and consider a neighborhood of the feasible homotopy path defined by

$$N_2(\beta; \overline{\mu}^{(k)}) = \left\{\mathbf{w} \in \mathcal{F}^0 : \left\|\left(\mathbf{XV} - \overline{\mu}^{(k)}\mathbf{X}_0\mathbf{V}_0\right)\mathbf{e}\right\|_2 \leq \beta\,\overline{\mu}^{(k)} \right.$$
$$\left. \text{and} \quad \mathbf{w} \in \mathcal{H}\left(\overline{\mathbf{w}}^{(k)}\right)\right\}, \tag{10.21}$$

where $\beta \in (0, 1)$, $\overline{\mu}^{(k)}$ is defined by (10.20), $\mathbf{X} = \mathrm{diag}[x_1, \ldots, x_n]$, $\mathbf{V} = \mathrm{diag}[v_1, \ldots, v_n]$, $\mathbf{X}_0$ and $\mathbf{V}_0$ are defined after (10.2), and $\mathcal{F}^0$ denotes the interior of the primal–dual feasible polytope. This neighborhood is closely related to the merit function (10.18). It measures distance as the norm of the residuals relative to the target duality gap $\overline{\mu}^{(k)}$, where the latter quantity would be the same if computed at any other point in the hyperplane.

If the point $\overline{\mathbf{w}}^{(k)}$ and its associated hyperplane $\mathcal{H}(\overline{\mathbf{w}}^{(k)})$ are varied, one obtains a *conical* neighborhood of the (lifted) feasible homotopy path as follows:

$$N_2(\beta) = \left\{ (\mathbf{x}, \pi, \mathbf{v}) \in \mathcal{F}^0 : \|\mathbf{XVe} - \mu \mathbf{X}_0 \mathbf{V}_0 \mathbf{e}\|_2 \leq \beta\mu, \right.$$

$$\left. \text{where } \mu = \frac{\mathbf{x}^T \mathbf{v}}{\left(\mathbf{x}^{(0)}\right)^T \mathbf{v}^{(0)}} \right\}. \qquad (10.22)$$

For the central path, this simplifies to

$$N_2(\beta) = \left\{ (\mathbf{x}, , \pi, \mathbf{v}) \in \mathcal{F}^0 : \|\mathbf{XVe} - \mu \mathbf{e}\|_2 \leq \beta\mu \text{ where } \mu = \frac{\mathbf{x}^T \mathbf{v}}{n} \right\}. \qquad (10.23)$$

Various other neighborhoods have been employed in the literature, for example, the neighborhood obtained by using the ∞-norm in place of the Euclidean norm in (10.22). Neighborhoods of lifted homotopy paths, in particular, neighborhoods of the central path, play an important role in the complexity analysis of associated interior-point algorithms.

Extensions

The foregoing considerations can be appropriately enlarged to the case of *infeasible* iterates. For example, consider the Newton corrector phase in the illustrative Algorithm EP/NC/R of Section 10.3. For reasons discussed immediately after expressions (10.16)–(10.17), the same merit function as in the feasible case, namely (10.18), can be used. Also, after iteration superscripts are explicitly attached, it follows directly from these two expressions that

$$\left(\overline{\mathbf{x}}^{(k)}\right)^T \overline{\mathbf{v}}^{(k)} = \left(\mathbf{c} + \mu^{(k)} \mathbf{q}_{v^{(0)}}\right)^T \overline{\mathbf{x}}^{(k)} - \left(\mathbf{b} + \mu^{(k)} \mathbf{q}_{x^{(0)}}\right)^T \overline{\pi}^{(k)}. \qquad (10.24)$$

If the associated target value $\overline{\mu}^{(k)} = \left(\overline{\mathbf{x}}^{(k)}\right)^T \overline{\mathbf{v}}^{(k)} / \left(\mathbf{x}^{(0)}\right)^T \mathbf{v}^{(0)}$ were used in a horizontal Newton corrector iteration started from $\overline{\mathbf{w}}^{(k)}$, then the iterates would lie in a hyperplane in primal–dual space that passes through $\overline{\mathbf{w}}^{(k)}$ and has the following normal vector:

$$\left(\left(\mathbf{c} + \mu^{(k)} \mathbf{q}_{v^{(0)}}\right), -\left(\mathbf{b} + \mu^{(k)} \mathbf{q}_{x^{(0)}}\right), \mathbf{0} \right).$$

Again a neighborhood analogous to (10.21) can be defined.

When the starting iterate for the Newton corrector is infeasible for the first two equations of the homotopy, then merit functions and neighborhoods must be suitably generalized. (This would occur in the foregoing example when a vertical predictor is substituted for the Euler predictor.)

These generalizations will not be pursued here, but further detail can be found in some of the relevant references cited in the next subsection.

10.4.4 Complexity Results

Theoretical algorithms based on the strategies summarized in Figure 10.2 and the measures and targets discussed in Section 10.4.3 have been extensively analyzed in the literature. We now give a short review[4] of some known complexity results.

Central path-following by Newton's method is based on *Strategy 1*. An embedding algorithm of this type, which is initiated from a suitably chosen feasible point, maintains feasible iterates, and uses a *single* Newton corrector step at each iteration, is shown to have polynomial complexity in Kojima, Mizuno, and Yoshise [1989]. Central path-following by Newton's method from an *infeasible* initial point—*Strategy 1* with $\mathbf{w}^{(0)} = (\mathbf{x}^{(0)}, \pi^{(0)}, \mathbf{v}^{(0)})$ and subsequent iterates permitted to be infeasible—is proposed in Lustig [1991], thereby extending the foregoing Kojima–Mizuno–Yoshise [1989] algorithm. A general *globally convergent* algorithm of this type is given in Kojima, Megiddo, and Mizuno [1993], and a particular version is shown to have polynomial complexity in Mizuno [1994]. A related result can be found in Zhang [1994].

An embedding algorithm based on *Strategy 2*, which is initiated from a feasible, interior point $\mathbf{w}^{(0)}$, maintains feasible iterates, and uses a single Newton corrector step at each iteration, is shown to have polynomial complexity in Nazareth [1991] using complexity analysis proof techniques developed by Renegar and Shub [1992]. The homotopy parameter is reduced at each vertical-predictor step according to the following rule:

$$\mu^{(k+1)} = \left(1 - \frac{1}{40\sqrt{n}\sqrt{\|\mathbf{W}_0\|_2 \|(\mathbf{W}_0)^{-1}\|_2}}\right) \mu^{(k)},$$

where $\mathbf{W}_0 \equiv \mathbf{X}_0 \mathbf{V}_0$ and $\|\cdot\|_2$ denotes the largest element on the diagonal of $\mathbf{W}_0$; i.e., its spectral norm. Note that the reduction in the parameter depends on the *condition number* (10.3) associated with the homotopy system (10.2). All subsequent iterates generated by the algorithm remain feasible, and the duality gap at each iteration is bounded as follows:

$$\left(\overline{\mathbf{x}}^{(k)}\right)^T \overline{\mathbf{v}}^{(k)} = \mathbf{c}^T \overline{\mathbf{x}}\left(\mu^{(k)}\right) - \mathbf{b}^T \overline{\pi}\left(\mu^{(k)}\right) \le 2\mu^{(k)} n \|\mathbf{W}_0\|_2.$$

The associated target value on the path is (10.20) and is thus bounded by

$$\overline{\mu}^{(k)} \le \frac{2\mu^{(k)} n \|\mathbf{W}_0\|_2}{\left(\mathbf{x}^{(0)}\right)^T \mathbf{v}^{(0)}}.$$

---

[4]This is by no means a comprehesive survey of the available literature, and many important contributions are not mentioned.

The duality gap and the target value can be reduced to any fixed prespecified accuracy in $O(\sqrt{n})$ iterations. Related results are given in Tseng [1992]. An algorithm based on *Strategy 2* that follows a path lying *outside* the feasible polytope is given by Kojima, Megiddo, and Noma [1991].

Strategy 3 is the traditional EN path-following strategy. Central path-following algorithms along these lines are developed and analyzed in Sonnevend, Stoer, and Zhao [1990], [1991] for the case of feasible starting points and iterates, with one or more corrector steps being used to return to a neighborhood of the central path. These authors also mention the more general "weighted path" case, again for a feasible starting point, but do not pursue its details.

Strategy 4 is implemented in the numerical illustration of Section 10.3 and corresponds to general homotopy path-following with restarts. No convergence or complexity results are available for this algorithm, but the computational results given in Section 10.3 and, in more detail, in Nazareth [1996b] provide some useful information about its performance on realistic test problems.

A polynomial-time Euler predictor/Newton corrector algorithm based on *Strategy 5* is given by Mizuno, Todd, and Ye [1993]. This algorithm is initiated and remains within the feasible region, and it uses a *single* corrector step at each iteration. An adaptive choice is made for the predictor step length and the homotopy parameter so as to remain within the central-path neighborhood defined by (10.23). As noted in Section 10.2.2, the Euler predictor for the feasible case defines the primal–dual affine-scaling direction, and its primal and dual components are directions of descent/ascent for the primal/dual objective vectors, respectively. Thus the duality gap decreases monotonically with a step along the Euler predictor. The target values at consecutive iterations decrease monotonically as follows:

$$\overline{\mu}^{(k+1)} \leq \left(1 - 8^{-0.25} n^{-0.5}\right)\overline{\mu}^{(k)}.$$

Finally, consider *Strategy 6* of Figure 10.2. Kojima, Megiddo, and Mizuno [1993] distinguish their *Strategy 1* path-following approach from the *Strategy 2* path-following approach of Kojima, Megiddo, and Noma [1991], both of which are mentioned in the discussion above, as follows:[5]

> The trajectory traced by their algorithm runs *outside* the feasible region, but the central trajectory traced by our algorithm runs through the interior of the feasible region.

In the same vein, *Strategy 6* should be distinguished from the central-path-following *Strategy 5*. Tseng [1995] gives a very accessible proof of polynomial complexity for an Euler predictor/Newton corrector approach based on *Strategy 6*, which is started on (or suitably near) the central ray and follows

---

[5] Italics ours.

a central-ray-initiated interior path as defined in Section 10.1.1. After an Euler predictor step, only a *single* Newton corrector step is performed at each iteration. The homotopy parameter and the infeasibility at successive iterations are explicitly reduced as follows:

$$\mu^{(k+1)} = \left(1 - \theta^{(k)}\right)\left(1 - \gamma^{(k)}\right)\mu^{(k)}$$

and

$$\begin{bmatrix} \mathbf{q}_{x^{(k+1)}} \\ \mathbf{q}_{v^{(k+1)}} \end{bmatrix} = \mu^{(k+1)} \begin{bmatrix} \mathbf{q}_{x^{(0)}} \\ \mathbf{q}_{v^{(0)}} \end{bmatrix},$$

where $\theta^k \in [0, 1]$ is a step length along the predictor direction that keeps the iterate within a neighborhood of the central-ray-initiated path ($\theta^k = 0$ is permitted), and $\gamma^k \in (0, 1)$ is a quantity associated with the corrector, for which a lower bound that depends on both n and the starting point ensures polynomial complexity. The other quantities define the current and initial infeasibilities analogously to their definition in earlier sections.

In summary and referring to Figure 10.1 for purposes of illustration, we see that complexity results are known for the central path C, the feasible homotopy path F, and the central-ray-initiated and optimally conditioned paths Ia and Ib. When the initiating point $\left(\mathbf{x}^{(0)}, \pi^{(0)}, \mathbf{v}^{(0)}\right)$, where $\mathbf{x}^{(0)} > \mathbf{0}$ and $\mathbf{v}^{(0)} > \mathbf{0}$, defines a general interior homotopy path as in path G in Figure 10.1, it is possible to extend the proof of Tseng [1995] to obtain a polynomial complexity bound that depends also on the condition number of $\mathbf{W}_0$. Results for noninterior paths—Section 10.1.2 and path N of Figure 10.1—can also be derived from the results mentioned in the present section through the use of the transformation (10.8). In this case the complexity result will depend on the condition number of the system, the "weights" ω_i and γ_i, and the distance of the starting point from feasibility. A detailed complexity analysis of algorithms based on paths illustrated by G and N of Figure 10.1 is an open topic that deserves further investigation.

10.5 Mehrotra's Implementation

Mehrotra [1991], [1992] achieved an *implementational breakthrough* via a clever synthesis and extension of bipath-following techniques discussed in Sections 10.2 and 10.4. In particular, his approach incorporated the second-order corrector of Section 10.2.3 into the search direction, utilized efficient heuristics for defining targets on the central path, used different steps in primal and dual space, and reorganized the computational linear algebra, namely, the solution of the large-scale systems of linear equations that define the search direction at each iteration, in order to enhance overall efficiency.

We introduce these ideas in stages, beginning with use of the second-order corrector of Section 10.2.3. Let us resume the discussion in the introductory paragraph of Section 10.3. A refinement of the basic Euler predictor with restarts strategy mentioned there is to supplement the Euler predictor (10.9), which linearizes and thus approximates the path to first order, by employing the second-order corrector (10.12) to now define a *quadratic approximation* to the path being followed. The algorithm below, which is due to Zhang and Zhang [1995], forms this quadratic approximation to *a homotopy path that emenates from the current iterate* $(\overline{\mathbf{x}}^{(k)}, \overline{\pi}^{(k)}, \overline{\mathbf{v}}^{(k)})$, and whose associated homotopy system is defined in an analogous manner to (10.2). In particular, the linear systems defining the algorithm's Euler predictor (10.25) and second-order corrector (10.26) are obtained from the systems of linear equations (10.9) and (10.12), respectively, with $(\mathbf{x}^{(0)}, \pi^{(0)}, \mathbf{v}^{(0)})$ replaced by $(\overline{\mathbf{x}}^{(k)}, \overline{\pi}^{(k)}, \overline{\mathbf{v}}^{(k)})$:

Algorithm EP/SOC/R: Given a starting point $(\overline{\mathbf{x}}^{(0)} > \mathbf{0}, \overline{\pi}^{(0)}, \overline{\mathbf{v}}^{(0)} > \mathbf{0})$, for $k = 0, 1, \ldots,$ do the following:

1. Solve the following set of linear equations for the Euler predictor direction $\mathbf{p}_{\overline{w}^{(k)}} = (\mathbf{p}_{\overline{x}^{(k)}}, \mathbf{p}_{\overline{\pi}^{(k)}}, \mathbf{p}_{\overline{v}^{(k)}})$:

$$\begin{aligned} \mathbf{A}\mathbf{p}_{\overline{x}^{(k)}} &= -\mathbf{q}_{\overline{x}^{(k)}}, \\ \mathbf{A}^T\mathbf{p}_{\overline{\pi}^{(k)}} + \mathbf{p}_{\overline{v}^{(k)}} &= -\mathbf{q}_{\overline{v}^{(k)}}, \\ \overline{\mathbf{X}}_k\mathbf{p}_{\overline{v}^{(k)}} + \overline{\mathbf{V}}_k\mathbf{p}_{\overline{x}^{(k)}} &= -\overline{\mathbf{X}}_k\overline{\mathbf{V}}_k\mathbf{e}, \end{aligned} \tag{10.25}$$

   where $\overline{\mathbf{X}}_k$ and $\overline{\mathbf{V}}_k$ are diagonal matrices defined by the components of $\overline{\mathbf{x}}^{(k)}$ and $\overline{\mathbf{v}}^{(k)}$, respectively, $\mathbf{q}_{\overline{x}^{(k)}} = \mathbf{A}\overline{\mathbf{x}}^{(k)} - \mathbf{b}$, and $\mathbf{q}_{\overline{v}^{(k)}} = \mathbf{A}^T\overline{\pi}^{(k)} + \overline{\mathbf{v}}^{(k)} - \mathbf{c}$.

2. Solve the following system of linear equations for the second-order corrector direction $\mathbf{s}_{\overline{w}^{(k)}} = (\mathbf{s}_{\overline{x}^{(k)}}, \mathbf{s}_{\overline{\pi}^{(k)}}, \mathbf{s}_{\overline{v}^{(k)}})$:

$$\begin{aligned} \mathbf{A}\mathbf{s}_{\overline{x}^{(k)}} &= \mathbf{0}, \\ \mathbf{A}^T\mathbf{s}_{\overline{\pi}^{(k)}} + \mathbf{s}_{\overline{v}^{(k)}} &= \mathbf{0}, \\ \overline{\mathbf{X}}_k\mathbf{s}_{\overline{v}^{(k)}} + \overline{\mathbf{V}}_k\mathbf{s}_{\overline{x}^{(k)}} &= -2\mathbf{p}_{\overline{x}^{(k)}}\mathbf{p}_{\overline{v}^{(k)}}, \end{aligned} \tag{10.26}$$

   where in the last set of equations of (10.26), the components of $\mathbf{p}_{\overline{x}^{(k)}}$ and $\mathbf{p}_{\overline{v}^{(k)}}$, which are obtained from (10.25), are multiplied pairwise.

3. Choose a step length $t^{(k)} > 0$ such that

$$\overline{\mathbf{w}}^{(k+1)} = \overline{\mathbf{w}}^{(k)} + t^{(k)}\mathbf{p}_{\overline{w}^{(k)}} + \tfrac{1}{2}\left(t^{(k)}\right)^2\mathbf{s}_{\overline{w}^{(k)}}$$

   remains positive in its $\overline{\mathbf{x}}^{(k+1)}$ and $\overline{\mathbf{v}}^{(k+1)}$ components.

   In a simpler version of Step 3, a direction-finding procedure (DfP) can be substituted for the quadratic approximation, by making a particular choice for $t^{(k)}$. Typically, choose $t^{(k)} = 1$, yielding a search direction as the sum of the Euler predictor and second-order corrector directions. Then take a step along this search direction, subject to remaining in the positive orthant with respect to the $\mathbf{x}$ and $\mathbf{v}$ variables.

Note that the homotopy path used to compute the predictor and second-order corrector is restarted, in effect, at each iteration. Like Algorithm EP/NC/R of Section 10.3, Algorithm EP/SOC/R is not known to be convergent. We now enhance the latter by utilizing a bipath-following strategy. In addition to the homotopy path used in the algorithm that emenates from the current iterate, an original guiding path is retained. *In the remainder of the discussion in this section, let us take the latter to be the central path.*

Referring back to the building blocks of Section 10.2 from which algorithms can be composed, let us consider the Newton corrector obtained from (10.13) at the point $\overline{\mathbf{w}}^{(k)}$ with associated parameter value $\mu^{(k)}$. Because the central path is being followed, $\mathbf{q}_{x^{(0)}} = \mathbf{0}$ and $\mathbf{q}_{v^{(0)}} = \mathbf{0}$, and thus the residuals, simplify in these equations. In the new notation, where iteration superscripts are attached, the equations defining the Newton corrector are as follows:

$$\begin{aligned} \mathbf{A}\mathbf{n}_{\overline{x}^{(k)}} &= -\mathbf{q}_{\overline{x}^{(k)}}, \\ \mathbf{A}^T\mathbf{n}_{\overline{\pi}^{(k)}} + \mathbf{n}_{\overline{v}^{(k)}} &= -\mathbf{q}_{\overline{v}^{(k)}}, \\ \overline{\mathbf{V}}_k\mathbf{n}_{\overline{x}^{(k)}} + \overline{\mathbf{X}}_k\mathbf{n}_{\overline{v}^{(k)}} &= -\overline{\mathbf{X}}_k\overline{\mathbf{V}}_k\mathbf{e} + \mu^{(k)}\mathbf{e}. \end{aligned} \tag{10.27}$$

From the relationship (10.15), the Newton corrector at $\overline{\mathbf{w}}^{(k)}$ is the sum of the Euler predictor computed for the homotopy restarted at $\overline{\mathbf{w}}^{(k)}$, namely, (10.25) and the centering component derived from (10.14) and weighted by $\mu^{(k)}$. This centering component, again in the present notation (with iteration numbers attached) and simplified for the central path, is as follows:

$$\begin{aligned} \mathbf{A}\mathbf{c}_{\overline{x}^{(k)}} &= \mathbf{0}, \\ \mathbf{A}^T\mathbf{c}_{\overline{\pi}^{(k)}} + \mathbf{c}_{\overline{v}^{(k)}} &= \mathbf{0}, \\ \overline{\mathbf{V}}_k\mathbf{c}_{\overline{x}^{(k)}} + \overline{\mathbf{X}}_k\mathbf{c}_{\overline{v}^{(k)}} &= \mathbf{e}. \end{aligned} \tag{10.28}$$

Motivated by Algorithm EP/SOC/R and the direction-finding procedure (DfP) that can be used within Step 3, the Newton corrector direction (10.27), in particular, the part that corresponds to the Euler predictor, is augmented by adding to it the second-order corrector obtained from (10.26). This yields the DfP employed within Mehrotra-type path-following approaches. Its search direction, stated mathematically, is the sum of the Euler predictor direction for the interior path restarted at $\overline{\mathbf{w}}^{(k)}$, the corresponding second-order corrector direction for this path, and the centering component direction for the central guiding path, where the last direction is weighted by $\mu^{(k)}$. Let us first consider the choice of $\mu^{(k)}$, and then the effective computation of the search direction.

Mehrotra [1992] proposed an adaptive heuristic technique for choosing the weight $\mu^{(k)}$ as follows:

$$\mu^{(k)} = \sigma^{(k)}\overline{\mu}^{(k)}, \tag{10.29}$$

where $\overline{\mu}^{(k)} = \left(\overline{\mathbf{x}}^{(k)}\right)^T \overline{\mathbf{v}}^{(k)}/n$ is the target value (10.20) on the central path, for which the denominator $\left(\mathbf{x}^{(0)}\right)^T \mathbf{v}^{(0)}$ equals n. The parameter $\sigma^{(k)} \in (0,1]$ is chosen adaptively at each iteration, using a heuristic based on the amount of progress that could be made along the pure Euler predictor (10.25). Recall from the itemized discussion at the end of Section 10.2.2 that this direction is equivalent to the primal–dual affine-scaling direction defined at the current iterate $\overline{\mathbf{w}}^{(k)}$. The heuristic proposed by Mehrotra [1992] is as follows. Compute steps along the components of the primal–dual affine-scaling direction in the usual way, for example, analogously to the computation of these quantities described in Chapters 7 and 8 for the feasible case. (This allows for different step lengths in the primal and dual spaces and thus permits greater potential progress.) Suppose the primal and dual points obtained using these step lengths are the components of the vector $\hat{\mathbf{w}}^{(k)} = \left(\hat{\mathbf{x}}^{(k)}, \hat{\pi}^{(k)}, \hat{\mathbf{v}}^{(k)}\right)$. A new target value $\hat{\mu}^{(k)} = \left(\hat{\mathbf{x}}^{(k)}\right)^T \hat{\mathbf{v}}^{(k)}/n$ is derived from them. Then let $\sigma^{(k)} = \left(\hat{\mu}^{(k)}/\overline{\mu}^{(k)}\right)^r$, where, typically, $r = 2$ or 3. This heuristic choice has been shown to work well in practice. The Newton corrector using (10.29) targets a point on a shifted hyperplane as described in Section 10.4.3. Also, it is useful to define

$$\sigma^{(k)}\overline{\mu}^{(k)} = \left(\overline{\mathbf{x}}^{(k)}\right)^T \overline{\mathbf{v}}^{(k)}/\rho, \tag{10.30}$$

and $\sigma^k \in (0,1]$ implies $\rho \geq n$. The above equation (10.30) will he helpful later in seeing the connection with Karmarkar-like potential-function developments of Chapter 12.

Finally, the computational linear algebra is organized efficiently in the Mehrotra implementation by exploiting the fact that the Euler predictor, the second-order corrector, and the centering component are obtained from systems of linear equations that differ only in their right-hand sides. Thus the matrix factorization used to compute the Euler predictor can be preserved and used to solve a second system for $\overline{\mathbf{c}}_{\overline{w}^{(k)}} = \left(\overline{\mathbf{c}}_{\overline{x}^{(k)}}, \overline{\mathbf{c}}_{\overline{\pi}^{(k)}}, \overline{\mathbf{c}}_{\overline{v}^{(k)}}\right)$ as follows:

$$\begin{aligned} \mathbf{A}\overline{\mathbf{c}}_{\overline{x}^{(k)}} &= \mathbf{0}, \\ \mathbf{A}^T\overline{\mathbf{c}}_{\overline{\pi}^{(k)}} + \overline{\mathbf{c}}_{\overline{v}^{(k)}} &= \mathbf{0}, \\ \overline{\mathbf{X}}_k\overline{\mathbf{c}}_{\overline{v}^{(k)}} + \overline{\mathbf{V}}_k\overline{\mathbf{c}}_{\overline{x}^{(k)}} &= -2\mathbf{p}_{\overline{x}^{(k)}}\mathbf{p}_{\overline{v}^{(k)}} + \sigma^{(k)}\overline{\mu}^{(k)}\mathbf{e}; \end{aligned} \tag{10.31}$$

i.e., instead of computing two separate directions—the second-order corrector and the weighted centering component—solve a single linear system, with a suitably chosen right-hand side as given in (10.31). Only one additional back-substitution operation is required to obtain the solution of this system, because the matrix factorization is already available. The solution $\overline{\mathbf{c}}_{\overline{w}^{(k)}}$ is the required sum of the second-order corrector $\mathbf{s}_{\overline{w}^{(k)}}$ given by (10.26) and the weighted centering component $\left(\sigma^{(k)}\overline{\mu}^{(k)}\right)\mathbf{c}_{\overline{w}^{(k)}}$ derived from (10.28), which we call the *centered second-order corrector.*

The search direction employed in the Mehrotra-type DfP is the sum of the Euler predictor $\mathbf{p}_{\overline{w}^{(k)}}$ given by (10.25) and the centered second-order corrector $\overline{\mathbf{c}}_{\overline{w}^{(k)}}$ given by (10.31). A step is taken along this direction, subject to remaining in the positive orthant with respect to the $\mathbf{x}$ and $\mathbf{v}$ variables in the usual way. Again, greater progress can be obtained by allowing different step lengths along directions in primal and dual space. Note also that the techniques for *inexact* computation of direction vectors, along lines discussed in Chapter 7, can be adapted to the present setting. Details of this refinement are left to the reader.

Mehrotra [1992] used a potential function and a "fall-back" search direction in order to guarantee convergence of his algorithm. A theoretical version of the Mehrotra predictor–corrector approach was shown to be of polynomial-time complexity by Zhang and Zhang [1995]. In their approach, the algorithm was restored to the infeasible-interior-point, central-path-following tradition of Kojima, Megiddo, and Mizuno [1993], Mizuno [1994], and Zhang [1994] mentioned in Section 10.4.4. But instead of using the more standard Euler predictor and Newton corrector strategy, an Euler predictor and centered second-order corrector strategy was substituted.

The techniques described in this section (and variants on them) have been incorporated into a number of practical programs. See, in particular, Lustig, Marsden, and Shanno [1992], [1994a], Vanderbei [1995], Gondzio [1995], and Zhang [1995].

10.6 Summary

In the early days of the "Karmarkar Revolution," the term "interior" signified the interior of the feasible polytope. Other methods were termed "exterior." The terminology has since evolved, and "interior," nowadays, means the interior of the positive orthant with respect to the $\mathbf{x}$ and/or $\mathbf{v}$ variables; i.e., the terminology has evolved from *polytope-interior* to *orthant-interior.* Methods that do not maintain feasibility with respect to the equality constraints of the primal and dual linear programs are called *infeasible-interior.* If the $\mathbf{x} \geq \mathbf{0}$ and/or $\mathbf{v} \geq \mathbf{0}$ bounds are violated, they are called *noninterior.*[6]

From a conceptual standpoint, we have seen in this chapter that the homotopy approach lies at the foundation of interior-point methods for linear programming. It has found *fundamental new expression* within this specialized context, and these adaptations and developments, in turn, have

[6] And thus, if equality constraints were also violated, they would be called *infeasible-noninterior* methods—(un)naturally!

considerably reilluminated usage within the traditional general nonlinear equations setting.

Modern interior-point techniques are based on predictor–corrector, bipath-following strategies, where the guiding path is usually the central path. Many interesting refinements remain to be explored. For example:

- Alternative parameterizations, in particular, the use of two parameters simultaneously, can provide a conceptual foundation for taking different steps in the primal and dual spaces. For some early proposals along these lines, see Nazareth [1996b].
- Alternative guiding paths can be used in place of the central path in many algorithms, for example, in Mehrotra's algorithm of Section 10.5.
- Noninterior extensions described in Section 10.1.2 are an interesting avenue to pursue.
- Unification and integration of self-dual simplex homotopy techniques, which follow piecewise-linear paths, and primal–dual interior-point homotopy techniques, which follow smooth paths, could lead to useful enhancements of current, large-scale mathematical programming systems. For some further discussion, see Nazareth [1996b].

10.7 Notes

Sections 10.1–10.5: The material in this chapter is based on Nazareth [1996b], [1998].

11

Log-Barrier Transformations

When the connection between Karmakar's interior-point algorithm (Karmarkar [1984]) and classical logarithmic barrier algorithms (Frisch [1955], [1956], Fiacco and McCormick [1968]) was first observed in the mid-nineteen eighties, the latter techniques were considered to be outmoded. Thus, the initial inference, drawn amidst the considerable controversy surrounding Karmarkar's algorithm, was that the relationship shed a *negative* light on the new interior-point approach. However, as interior-point LP algorithms evolved and prospered, this view began to recede. Instead, logarithmic barrier methods themselves were resurrected and given fresh life. Like homotopy methods of the previous chapter, log-barrier methods found *fundamental new expression* within the specialized setting of large-scale linear programming.

In this chapter we give an a priori justification for logarithmic barrier transformations of a linear program and their relevance for interior-point algorithms. This approach is premised on the foundational global homotopy system (10.2) of Chapter 10 and the symmetry principle of Section 3.3, and it is presented in primal, dual, and primal–dual LP settings.

Logarithmic barrier transformations can be used to establish existence and uniqueness properties of the solution to the homotopy system (10.2), which is their primary role within the present monograph. They also serve to motivate the Karmarkar potential function approach of the next chapter. The use of logarithmic barrier transformations to formulate and implement interior-point *algorithms* for linear programming is considered only very briefly in the concluding section.

11.1 Derivation

Our starting premise is the homotopy system of equations (10.2) with $\mu > 0$. The Jacobian matrix of (10.2) is defined by (10.4), and it is not symmetric. Thus, by the symmetry principle of Section 3.3, a potential function for the mapping in (10.2) *cannot* exist. We set out to *reformulate* the system (10.2) in order to obtain an *equivalent system* whose Jacobian matrix is symmetric.

Two such reformulations are discussed, and associated potential functions are constructed. The connection with logarithmic barrier transformations of the linear programs (P) and (D) in (8.1) and implications for establishing properties of homotopy paths will then immediately become apparent.

In what follows, it will be convenient to make the definitions:

$$\begin{aligned} \mathbf{b}(\mu) &= \mathbf{b} + \mu \mathbf{q}_{x^{(0)}}, \\ \mathbf{c}(\mu) &= \mathbf{c} + \mu \mathbf{q}_{v^{(0)}}. \end{aligned} \tag{11.1}$$

11.1.1 *Primal*

Use the relation $\mathbf{v} = \mathbf{V}\mathbf{e} = \mu \mathbf{X}^{-1}(\mathbf{X}_0 \mathbf{V}_0)\mathbf{e}$, which is obtained from the third equation of (10.2), to eliminate the dual slack variables $\mathbf{v}$ from the second equation. Then reorder the remaining equations and variables to obtain the system

$$\begin{aligned} \mu \mathbf{X}^{-1}(\mathbf{X}_0 \mathbf{V}_0)\mathbf{e} + \mathbf{A}^T \pi - \mathbf{c}(\mu) &= \mathbf{0}, \\ \mathbf{A}\mathbf{x} - \mathbf{b}(\mu) &= \mathbf{0}, \end{aligned} \tag{11.2}$$

with $\mu > 0$ and $\mathbf{x} \in R^n_{++}$.

Define $\mathbf{z} = (\mathbf{x}, \pi)$ and identify the foregoing system (11.2) with the system of equations (3.1). Then, let $\bar{\mathbf{z}} = (\bar{\mathbf{x}}, \bar{\pi})$ for *any* fixed vectors $\bar{\mathbf{x}} > \mathbf{0}$ and $\bar{\pi}$, and use the construction of Section 3.3, along with some completely straightforward manipulation whose details are omitted, to obtain the following potential function, which is unique up to addition of an arbitrary scalar constant:

$$\begin{aligned} f_P(\mathbf{x}, \pi) = -\mathbf{c}(\mu)^T \mathbf{x} + \mu \sum_{i=1}^{n} \left(x_i^{(0)} v_i^{(0)} \right) \ln x_i \\ + \pi^T (\mathbf{A}\mathbf{x} - \mathbf{b}(\mu)). \end{aligned} \tag{11.3}$$

Note that this function is *independent of the particular choice* $\bar{\mathbf{x}}$ *and* $\bar{\pi}$.

A solution of (11.2), which is a reorganization of (10.2) of the previous chapter, corresponds to a point where the gradient of f_P vanishes. It is immediately evident that (11.3) is the Lagrangian function of the following weighted logarithmic barrier program:

$$\text{minimize } \mathbf{c}^T\mathbf{x} + \mu\mathbf{q}_{v^{(0)}}^T\mathbf{x} - \mu\sum_{i=1}^{n}\left(x_i^{(0)}v_i^{(0)}\right)\ln x_i,$$
$$\text{s.t. } \mathbf{A}\mathbf{x} = \mathbf{b} + \mu\mathbf{q}_{x^{(0)}}, \tag{11.4}$$

where $\mu > 0$ and $\mathbf{x} \in R^n_{++}$.

The foregoing *convex* program is very useful for establishing properties of paths in the previous chapter. Since the objective function in (11.4) is strictly convex and approaches $+\infty$ as any x_i approaches 0, (11.4) possesses a unique optimal solution that must lie in R^n_{++}. Hence the solution $(\mathbf{x}(\mu), \pi(\mu), \mathbf{v}(\mu))$ of the homotopy system (10.2) is unique for each $\mu > 0$ and lies in the interior of the nonnegative orthant with respect to the variables $\mathbf{x}$ and $\mathbf{v}$. Recall from (10.5) and the discussion preceding this expression that the "duality gap" associated with the initial point of the homotopy system (10.2) is $\left(\mathbf{x}^{(0)}\right)^T\mathbf{v}^{(0)}$ and that

$$\mu = \frac{\mathbf{x}(\mu)^T\mathbf{v}(\mu)}{\left(\mathbf{x}^{(0)}\right)^T\mathbf{v}^{(0)}}.$$

Recall also that μ could be given any positive and arbitrarily large value in (10.2); i.e., it need not be confined to the interval $(0, 1]$.

The foregoing observations and association between the homotopy system (10.2) and the convex program (11.4) imply the following (details of verification are straightforward and are left to the reader):

- The feasible regions of (P) and (D) cannot both be bounded simultaneously. Since we have assumed that (P) has a bounded feasible region, this cannot be true of (D).
- Given any number $\theta > 0$, there exist unique points $\mathbf{x}^{(0)} > \mathbf{0}$ and $\left(\pi^{(0)}, \mathbf{v}^{(0)} > \mathbf{0}\right)$ that are *feasible* for (P) and (D), respectively, i.e., their residual vectors satisfy $\mathbf{q}_{x^{(0)}} = \mathbf{0}$ and $\mathbf{q}_{v^{(0)}} = \mathbf{0}$, and whose components, in addition, satisfy

  $$x_i^{(0)}v_i^{(0)} = \theta \,\forall\, i.$$

  In particular, one can take $\theta = 1$.

  With the foregoing choices for $\mathbf{x}^{(0)}$ and $\left(\pi^{(0)}, \mathbf{v}^{(0)}\right)$, the convex program (11.4) simplifies considerably. As $\mu > 0$ is varied, its associated optimal primal and dual variables, or, equivalently, the solutions of the homotopy system (10.2), then define the unique *central path* $(\mathbf{x}(\mu), \pi(\mu), \mathbf{v}(\mu))$, $\mu > 0$, of the linear programs (P) and (D).

11.1.2 Dual

Reformulate the system (10.2) in a different way. Rewrite the third equation as $-\mu\mathbf{V}^{-1}(\mathbf{X}_0\mathbf{V}_0)\mathbf{e} + \mathbf{x} = \mathbf{0}$, and reorder the equations and variables to obtain the following system:

$$\begin{aligned} -\mu\mathbf{V}^{-1}(\mathbf{X}_0\mathbf{V}_0)\mathbf{e} + \mathbf{x} &= \mathbf{0}, \\ \mathbf{v} + \mathbf{A}^T\pi - \mathbf{c}(\mu) &= \mathbf{0}, \\ \mathbf{A}\mathbf{x} - \mathbf{b}(\mu) &= \mathbf{0}, \end{aligned} \tag{11.5}$$

with $\mu > 0$ and $\mathbf{x}, \mathbf{v} \in R^n_{++}$.

Define $\mathbf{z} = (\mathbf{v}, \mathbf{x}, \pi)$, and again observe that the system of equations corresponding to (11.5) has a symmetric Jacobian matrix. As in the previous subsection for the primal, form the potential function using the construction of Section 3.3. Straightforward manipulation, whose details are again omitted, gives

$$f_D(\mathbf{v}, \mathbf{x}, \pi) = -\mathbf{b}(\mu)^T\pi - \mu\sum_{i=1}^{n}\left(x_i^{(0)}v_i^{(0)}\right)\ln v_i + \mathbf{x}^T\left(\mathbf{A}^T\pi + \mathbf{v} - \mathbf{c}(\mu)\right). \tag{11.6}$$

A solution of (11.5), and hence (10.2), corresponds to a point where the gradient of f_D vanishes. Again, it is evident that (11.6) is the Lagrangian function of the following weighted logarithmic barrier convex program:

$$\begin{aligned} &\text{maximize } \mathbf{b}^T\pi + \mu\mathbf{q}^T_{x^{(0)}}\pi + \mu\sum_{i=1}^{n}\left(x_i^{(0)}v_i^{(0)}\right)\ \ln\ v_i, \\ &\text{s.t. } \mathbf{A}^T\pi + \mathbf{v} = \mathbf{c} + \mu\mathbf{q}_{v^{(0)}}, \end{aligned} \tag{11.7}$$

where $\mu > 0$ and $\mathbf{v} \in R^n_{++}$.

11.1.3 Primal–Dual

The foregoing development implies that (11.4) and (11.7) have the same solution. Therefore, we can seek to solve both problems *simultaneously*, yielding the following separable weighted logarithmic barrier convex program:

$$\begin{aligned} \min\left(\mathbf{c}^T\mathbf{x} - \mathbf{b}^T\pi\right) + \mu\left(\mathbf{q}^T_{v^{(0)}}\mathbf{x}\ -\ \mathbf{q}^T_{x^{(0)}}\pi\right) - \mu\sum_{i=1}^{n}\left(x_i^{(0)}v_i^{(0)}\right)\ \ln x_i v_i, \\ \text{s.t. } \mathbf{A}\mathbf{x} = \mathbf{b} + \mu\mathbf{q}_{x^{(0)}}, \\ \mathbf{A}^T\pi + \mathbf{v} = \mathbf{c} + \mu\mathbf{q}_{v^{(0)}}, \end{aligned} \tag{11.8}$$

where $\mu > 0$ and $\mathbf{x}, \mathbf{v} \in R^n_{++}$.

Observe that the optimal Lagrange multipliers associated with the two constraints of (11.8) must themselves be the optimal values of the π and $\mathbf{x}$ variables, respectively. This follows directly from the way in which (11.8) was formulated. Observe also that (11.8) can be obtained directly by making a nonstandard logarithmic barrier transformation of the nonnegativity constraints of the following *self-dual* linear program:

$$\begin{aligned} \text{minimize } & \mathbf{c}^T\mathbf{x} - \mathbf{b}^T\pi, \\ \text{s.t. } & \mathbf{A}\mathbf{x} = \mathbf{b}, \\ & \mathbf{A}^T\pi + \mathbf{v} = \mathbf{c}, \\ & \mathbf{x} \geq \mathbf{0}, \mathbf{v} \geq \mathbf{0}. \end{aligned} \tag{11.9}$$

Use the constraints of (11.8) to reexpress this convex program more conveniently as follows:

$$\begin{aligned} \text{minimize } & \mathbf{x}^T\mathbf{v} - \mu \sum_{i=1}^{n} \left(x_i^{(0)} v_i^{(0)}\right) \ln x_i v_i, \\ \text{s.t. } \mathbf{A}\mathbf{x} &= \mathbf{b} + \mu \mathbf{q}_{x^{(0)}}, \\ \mathbf{A}^T\pi + \mathbf{v} &= \mathbf{c} + \mu \mathbf{q}_{v^{(0)}}, \end{aligned} \tag{11.10}$$

where $\mu > 0$ and $\mathbf{x}$, $\mathbf{v} \in R^n_{++}$. It will be used in this form to motivate primal–dual Karmarkar-like potentials in the next chapter.

11.2 Special Cases

Suppose the points $\mathbf{x}^{(0)}$ and $\left(\pi^{(0)}, v^{(0)}\right)$ lie on the central path; i.e., $\mathbf{q}_{x^{(0)}} = \mathbf{0}$, $\mathbf{q}_{v^{(0)}} = \mathbf{0}$, and $x_i^{(0)} v_i^{(0)} = 1 \, \forall i$; see the last item of Section 11.1.1. Then (11.4) and (11.7) correspond, respectively, to well-known *unweighted* logarithmic barrier transformations of the original primal and dual linear programs. Again, when components of $\left(\mathbf{x}^{(0)}, \pi^{(0)}, \mathbf{v}^{(0)}\right)$ are feasible, but these points lies off the central path, then (11.4) and (11.7) correspond, respectively, to *weighted* logarithmic barrier transformations.

An interesting case arises when the points $\mathbf{x}^{(0)}$ and/or $\left(\pi^{(0)}, \mathbf{v}^{(0)}\right)$ are permitted to be *infeasible* and one makes the simplest of possible choices, namely, $\mathbf{x}^{(0)} = \mathbf{e}$, $\pi^{(0)} = \mathbf{0}$, and $\mathbf{v}^{(0)} = \mathbf{e}$. The corresponding primal, dual, and primal–dual logarithmic barrier transformations of the original linear program are then as follows below.

11.2.1 Primal

$$\begin{aligned} \text{minimize } & (1-\mu)\mathbf{c}^T\mathbf{x} + \mu \mathbf{e}^T\mathbf{x} - \mu \sum_{i=1}^{n} \ln x_i, \\ \text{s.t. } & \mathbf{A}\mathbf{x} = (1-\mu)\mathbf{b} + \mu \mathbf{A}\mathbf{e}. \end{aligned} \tag{11.11}$$

where $\mu > 0$ and $\mathbf{x} \in R^n_{++}$.

11.2.2 *Dual*

Trivial reformulation of the objective function of (11.7) and dropping of fixed terms involving $\mu, \mathbf{c}$, and $\mathbf{e}$ yields

$$\begin{aligned} \text{minimize } & -(1-\mu)\mathbf{b}^T\pi + \mu\mathbf{e}^T\mathbf{v} - \mu\sum_{i=1}^{n} \ln v_i, \\ \text{s.t. } & \mathbf{A}^T\pi + \mathbf{v} = (1-\mu)\mathbf{c} + \mu\mathbf{e}. \end{aligned} \tag{11.12}$$

where $\mu > 0$ and $\mathbf{v} \in R^n_{++}$.

11.2.3 *Primal–Dual*

$$\begin{aligned} \text{minimize } & \mathbf{x}^T\mathbf{v} - \mu\sum_{i=1}^{n} \ln x_i v_i, \\ \text{s.t. } & \mathbf{A}\mathbf{x} = (1-\mu)\mathbf{b} + \mu\mathbf{A}\mathbf{e}, \\ & \mathbf{A}^T\pi + \mathbf{v} = (1-\mu)\mathbf{c} + \mu\mathbf{e}. \end{aligned} \tag{11.13}$$

where $\mu > 0$ and $\mathbf{x}, \mathbf{v} \in R^n_{++}$.

11.3 Discussion

Although not pursued here, unweighted logarithmic barrier transformations of a linear program can be made a cornerstone for the formulation, analysis, and implementation of interior-point algorithms; see, for example, Frisch [1956], Gill et al. [1986], Lustig et al. [1994b].

The more general log-barrier transformations of this chapter for the *infeasible, weighted* case can also be made the basis for interior-point algorithms and their implementations. The derivation in Section 11.1 provides a rationale for the particular choice of weights $x_i^{(0)}v_i^{(0)}, i = 1, \ldots, n$, associated with a feasible interior solution $(\mathbf{x}^{(0)}, \pi^{(0)}, \mathbf{v}^{(0)})$ of (10.1). It also yields some novel and potentially useful log-barrier transformations for the case when $\mathbf{x}^{(0)} > \mathbf{0}, \pi^{(0)}, \mathbf{v}^{(0)} > \mathbf{0}$, and the components of $(\mathbf{x}^{(0)}, \pi^{(0)}, \mathbf{v}^{(0)})$ are not feasible for the equality constraints of (P) and (D). For related work, see R. Polyak [1992], Freund [1996].

Many interior-point algorithmic variants arise from relaxing nonnegative constraints on dual variables in the setting of (11.4) or (11.11), and on primal variables in the setting of (11.7) or (11.12). Explicit linkages between primal variables and dual slack variables derived from the third equation of (10.2) can also be imposed in problems (11.8) or (11.13).

For the *noninterior* case discussed in Section 10.1.2, reformulations analogous to earlier sections of this chapter can be performed, yielding *shifted* log-barrier transformations. This is an interesting area that deserves further investigation. For related work, see, for example, Chen and Harker [1993] and Burke and Xu [2000].

11.4 Notes

Section 11.0: The connection between Karmarkar's algorithm and log-barrier techniques was investigated, in detail, by Gill, Murray, Saunders, Tomlin, and Wright [1986].

Sections 11.1–11.2: This material is derived from Nazareth [1997a]. For a related discussion on the existence of potentials for the system (10.2), see Vanderbei [1994]. For other types of reformulation of (10.2) with *nonsymmetric* Jacobian matrices and resulting algorithms, see Nazareth [1994d], [1998].

Section 11.3: For the definitive treatment of self-concordant log-barrier-based interior methods, see the incandescent monograph of Nesterov and Nemirovsky [1994] and the many references given therein.

12

Karmarkar Potentials and Algorithms

Karmarkar's algorithmic breakthrough has been characterized by Anstreicher [1996] as follows:[1]

> Potential reduction algorithms have a distinguished role in the area of interior point methods for mathematical programming. Karmarkar's algorithm for linear programming, whose announcement in 1984 initiated a torrent of research into interior point methods, used three key ingredients: a nonstandard linear programming formulation, projective transformations, and a potential function with which to measure progress of the algorithm. It was quickly shown that the nonstandard formulation could be avoided, and eventually algorithms were developed that eliminated the projective transformations, but retained use of the potential function. *It is then fair to say that the only really essential element of Karmarkar's analysis was the potential function....* In the classical optimization literature, potential reduction algorithms are most closely related to Huard's "method of centers"; see also Fiacco and McCormick [1968, Section 7.2]. However, Karmarkar's use of a potential function to facilitate a *complexity*, as opposed to a *convergence* analysis, was completely novel.

Karmarkar's 1984 formulation of a potential function was for a linear program in *primal* (or equivalently dual) form, and it has been supplanted

[1] Italics ours.

by developments in a primal–dual setting. Motivation for the latter comes directly from the primal–dual log-barrier potentials and associated transformations of a linear program discussed in Chapter 11.

In this chapter, unweighted and weighted Karmarkar-like primal–dual potential functions are presented. Then a specific primal–dual potential-reduction LP algorithm is formulated, and its complexity properties are outlined.

In the concluding section, the interconnections between primal–dual affine scaling, potential-reduction, and path-following algorithms are briefly discussed, along with implications for the effective implementation of potential-reduction algorithms.

12.1 Derivation

12.1.1 Notation

We will no longer need to use a bar to distinguish between iterates of an algorithm that lie on and iterates that lie off a homotopy path as in Chapter 10. Thus we revert to the simpler notation of Chapters 7 and 8, and henceforth denote a current iterate by $\mathbf{w}^{(k)} = \left(\mathbf{x}^{(k)}, \pi^{(k)}, \mathbf{v}^{(k)}\right)$.

12.1.2 Unweighted Case

Consider the following function:

$$\Phi_\rho(\mathbf{x}, \pi, \mathbf{v}) = \rho \ln \mathbf{x}^T \mathbf{v} - \sum_{i=1}^{n} \ln x_i v_i, \tag{12.1}$$

where $\rho \geq n$, and $\mathbf{x} > \mathbf{0}$ and $\mathbf{v} > \mathbf{0}$ are confined to the *interiors* of the *feasible polytopes* of (P) and (D), respectively.

Because our aim is to employ the function (12.1) as a *potential*, the vector field defined by its gradient vector is of interest. Specifically, let us consider this vector at a feasible interior point $\mathbf{w}^{(k)} = \left(\mathbf{x}^{(k)} > \mathbf{0}, \pi^{(k)}, \mathbf{v}^{(k)} > \mathbf{0}\right)$, namely,

$$\nabla \Phi_\rho\left(\mathbf{x}^{(k)}, \pi^{(k)}, \mathbf{v}^{(k)}\right) = \begin{bmatrix} \nabla_x \Phi_\rho\left(\mathbf{x}^{(k)}, \pi^{(k)}, \mathbf{v}^{(k)}\right) \\ \nabla_\pi \Phi_\rho\left(\mathbf{x}^{(k)}, \pi^{(k)}, \mathbf{v}^{(k)}\right) \\ \nabla_v \Phi_\rho\left(\mathbf{x}^{(k)}, \pi^{(k)}, \mathbf{v}^{(k)}\right) \end{bmatrix}$$

$$= \begin{bmatrix} \dfrac{\rho}{\left(\mathbf{x}^{(k)}\right)^T \mathbf{v}^{(k)}} \mathbf{v}^{(k)} - \left[\mathbf{x}^{(k)}\right]^{-1} \\ \mathbf{0} \\ \dfrac{\rho}{\left(\mathbf{x}^{(k)}\right)^T \mathbf{v}^{(k)}} \mathbf{x}^{(k)} - \left[\mathbf{v}^{(k)}\right]^{-1} \end{bmatrix}, \tag{12.2}$$

where $\left[\mathbf{x}^{(k)}\right]^{-1}, [\mathbf{v}^{(k)}]^{-1}$ are n-vectors with components $\left(x_i^{(k)}\right)^{-1}$ and $\left(v_i^{(k)}\right)^{-1}$, respectively.

Now consider the log-barrier function in (11.10), which was obtained from the potentials (11.3) and (11.6) associated with reformulations of the homotopy system (10.2). Restrict attention to the homotopy system that defines the *central path*, so assume that $\left(\mathbf{x}^{(0)}, \pi^{(0)}, \mathbf{v}^{(0)}\right)$ is a feasible interior point for (P) and (D), $\mathbf{q}_{x^{(0)}} = \mathbf{0}$ and $\mathbf{q}_{v^{(0)}} = \mathbf{0}$, and the weights in the logarithmic term of (11.10) or, equivalently, the initiating point components in (10.2), satisfy $\left(x_i^{(0)} v_i^{(0)}\right) = 1 \, \forall i$; see also the last item in Section 11.1.1. Because $\mu > 0$ is an *exogenous* parameter; i.e., a constant within this convex program, let us simplify (11.10) using the foregoing assumptions and reexpress it as follows:

$$\text{minimize } f_{PD}(\mathbf{x}, \pi, \mathbf{v}) = \frac{\mathbf{x}^T \mathbf{v}}{\mu} - \sum_{i=1}^{n} \ln x_i v_i, \tag{12.3}$$

where $\mathbf{x}$ and $\mathbf{v}$ are confined to the interiors of the feasible polytopes of (P) and (D), respectively.

The gradient vector of f_{PD} at the point $\left(\mathbf{x}^{(k)}, \pi^{(k)}, \mathbf{v}^{(k)}\right)$ is as follows:

$$\nabla f_{PD}\left(\mathbf{x}^{(k)}, \pi^{(k)}, \mathbf{v}^{(k)}\right) = \begin{bmatrix} \frac{1}{\mu}\mathbf{v}^{(k)} - \left[\mathbf{x}^{(k)}\right]^{-1} \\ \mathbf{0} \\ \frac{1}{\mu}\mathbf{x}^{(k)} - [\mathbf{v}^{(k)}]^{-1} \end{bmatrix}. \tag{12.4}$$

Thus, the two functions Φ_ρ and f_{PD} will have the same gradient vectors and generate the same associated vector field when μ is chosen as follows:

$$\mu = \frac{\left(\mathbf{x}^{(k)}\right)^T \mathbf{v}^{(k)}}{\rho}. \tag{12.5}$$

Make the following definitions:

$$\mu^{(k)} = \frac{\left(\mathbf{x}^{(k)}\right)^T \mathbf{v}^{(k)}}{n}, \qquad \sigma = n/\rho. \tag{12.6}$$

Then

$$\mu = \sigma \mu^{(k)}, \qquad \sigma \in (0, 1]. \tag{12.7}$$

In the notation of the present chapter (Section 12.1.1), we see that (12.7) is precisely the choice (10.29)–(10.30). *Thus we can view the potential function (12.1) as making an endogenous or automatic choice for the parameter μ in the primal–dual log-barrier function (12.3) and for the associated target point on the central path.* In constrast to (10.29)–(10.30), the target point is *nonadaptive*, because ρ and σ are fixed quantities.

Let us now restrict attention to points $(\mathbf{x}, \pi, \mathbf{v})$ that lie on the central path; i.e., points that satisfy the condition

$$x_i v_i = \frac{\mathbf{x}^T\mathbf{v}}{n}, \qquad i = 1, \ldots, n,$$

which follows directly from $x_i v_i = \mu \geq 0,\ i = 1, \ldots, n$. Then taking the logarithm on each side and summing over n gives

$$\sum_{i=1}^{n} \ln x_i v_i = n \ln \mathbf{x}^T\mathbf{v} - n \ln n.$$

Thus the potential function Φ_ρ can be expressed as

$$\Phi_\rho(\mathbf{x}, \pi, \mathbf{v}) = (\rho - n) \ln \mathbf{x}^T\mathbf{v} + n \ln n,$$

and Φ_ρ is seen to be *constant on the central path* when $\rho = n$. The function Φ_ρ tends to $-\infty$ as $\mathbf{x}^T\mathbf{v} \downarrow 0$ whenever $\rho > n$.

12.1.3 Weighted Case

Similarly, given any feasible interior point $\left(\mathbf{x}^{(0)}, \pi^{(0)}, \mathbf{v}^{(0)}\right)$, the weighted logarithmic barrier program (11.10) motivates the following weighted primal–dual potential function, in a manner analogous to the formulation in the foregoing unweighted case:

$$\Phi_\rho^w(\mathbf{x}, \pi, \mathbf{v}) = \rho \ln \mathbf{x}^T\mathbf{v} - \sum_{i=1}^{n} (x_i^{(0)} v_i^{(0)}) \ln x_i v_i \tag{12.8}$$

where $\rho \geq (\mathbf{x}^{(0)})^T\mathbf{v}^{(0)}$ and $\mathbf{x} > \mathbf{0}$ and $\mathbf{v} > \mathbf{0}$ are confined to the *interiors* of the *feasible polytopes* of (P) and (D), respectively.

One can parallel the development in Section 12.1.2 to show that the function (12.8) is constant on the homotopy path defined by (10.2) when $\rho = (\mathbf{x}^{(0)})^T\mathbf{v}^{(0)} \equiv \omega^{(0)}$. It tends to $-\infty$ as $\mathbf{x}^T\mathbf{v} \downarrow 0$ whenever $\rho > \omega^{(0)}$.

The weighted potential (12.8) will have a relationship to the weighted barrier function that is analogous to the relationship described in Section 12.1.2 for the unweighted case. In particular, it provides an automatic (or endogenous) choice for μ and the associated target on the *homotopy* path. Details of this extension are left to the reader.

12.2 A Potential-Reduction Algorithm

Given any feasible interior iterate $\left(\mathbf{x}^{(k)}, \pi^{(k)}, \mathbf{v}^{(k)}\right)$, we formulate a Newton–Cauchy algorithm along lines that closely parallel the primal–dual affine-scaling formulation of Section 8.3, but now using the potential function in place of the objective functions of (P) and (D).

Thus, let us consider primal and dual direction-finding problems analogous to (8.11) and (8.12) as follows:

$$\min \nabla_x \Phi_\rho \left(\mathbf{x}^{(k)}, \pi^{(k)}, \mathbf{v}^{(k)}\right)^T \left(\mathbf{x} - \mathbf{x}^{(k)}\right) + \tfrac{1}{2}\left(\mathbf{x} - \mathbf{x}^{(k)}\right)^T \mathbf{D}_X^{-2}\left(\mathbf{x} - \mathbf{x}^{(k)}\right)$$
$$\text{s.t. } \mathbf{A}\mathbf{x} - \mathbf{b} = \mathbf{A}\left(\mathbf{x} - \mathbf{x}^{(k)}\right) = \mathbf{0} \tag{12.9}$$

and

$$\max - \left[\mathbf{0}^T, \nabla_v \Phi_\rho\left(\mathbf{x}^{(k)}, \pi^{(k)}, \mathbf{v}^{(k)}\right)^T\right] \left[\begin{array}{c} \left(\pi - \pi^{(k)}\right) \\ \left(\mathbf{v} - \mathbf{v}^{(k)}\right) \end{array}\right]$$
$$- \tfrac{1}{2}\left(\mathbf{v} - \mathbf{v}^{(k)}\right)^T \mathbf{D}_V^{-2}\left(\mathbf{v} - \mathbf{v}^{(k)}\right)$$
$$\text{s.t. } \mathbf{A}^T \pi + \mathbf{v} - \mathbf{c} = \mathbf{A}^T\left(\pi - \pi^{(k)}\right) + \left(\mathbf{v} - \mathbf{v}^{(k)}\right) = \mathbf{0}, \tag{12.10}$$

where, as in Section 8.3, $\mathbf{D}_X$ and $\mathbf{D}_V$ are diagonal matrices with positive diagonal elements. Comparing the above with (8.11) and (8.12) shows that the only difference lies in replacing the primal and dual objective vectors; i.e., the gradients of the objective functions of (P) and (D), with the gradients of the primal–dual potential function (12.1) with respect to the variables $\mathbf{x}$ and $\mathbf{v}$.

Make the following definitions:

$$\mathbf{d} = \left(\mathbf{x} - \mathbf{x}^{(k)}\right), \qquad \mathbf{h}_\pi = \left(\pi - \pi^{(k)}\right), \qquad \mathbf{h}_v = \left(\mathbf{v} - \mathbf{v}^{(k)}\right).$$

Then the Lagrangian optimality conditions for (12.9) are as follows:

$$\mathbf{A}\mathbf{d} = \mathbf{0},$$
$$\mathbf{D}_X^{-2}\mathbf{d} - \mathbf{A}^T\lambda = -\nabla_x \Phi_\rho(\mathbf{x}^{(k)}, \pi^{(k)}, \mathbf{v}^{(k)}), \tag{12.11}$$

where λ denotes the vector of Langrange multipliers associated with the equality constraints.

Let $\mathbf{z} = -\mathbf{A}^T\lambda$. Then, using (12.2), we can reexpress the optimality conditions as follows:

$$\mathbf{A}\mathbf{d} = \mathbf{0},$$
$$\mathbf{A}^T\lambda + \mathbf{z} = \mathbf{0},$$
$$\mathbf{D}_X^{-2}\mathbf{d} + \mathbf{z} = \frac{\rho}{\left(\mathbf{x}^{(k)}\right)^T \mathbf{v}^{(k)}}\mathbf{v}^{(k)} - \mathbf{X}_k^{-1}\mathbf{e}, \tag{12.12}$$

where $\mathbf{X}_k$ is a diagonal matrix with diagonal elements given by the components of $\mathbf{x}^{(k)}$.

Similarly, the Lagrangian optimality conditions for (12.10) can be stated as follows:

$$\mathbf{A}\mathbf{u} = \mathbf{0},$$
$$\mathbf{A}^T\mathbf{h}_\pi + \mathbf{h}_v = \mathbf{0},$$
$$\mathbf{u} + \mathbf{D}_V^{-2}\mathbf{h}_v = \frac{\rho}{\left(\mathbf{x}^{(k)}\right)^T \mathbf{v}^{(k)}}\mathbf{x}^{(k)} - \mathbf{V}_k^{-1}\mathbf{e}, \tag{12.13}$$

where $\mathbf{u}$ denotes a vector of Lagrange multipliers associated with the constraints of (12.10), and $\mathbf{V}_k$ is a diagonal matrix with diagonal elements given by the components of $\mathbf{v}^{(k)}$.

The above pair of Lagrangian optimality conditions become identical, in different variables, when one imposes the self-dual condition $\mathbf{D}_X = \mathbf{D}_V^{-1}$ and defines $\mathbf{D}_X$ by (8.17), namely,

$$\mathbf{D}_X = \mathrm{diag}\left[\sqrt{\left(x_1^{(k)}/v_1^{(k)}\right)}, \ldots, \sqrt{\left(x_n^{(k)}/v_n^{(k)}\right)}\right] = \mathbf{D}_V^{-1}. \qquad (12.14)$$

By analogy with (8.18) and (8.19), or by direct derivation from the Lagrangian optimality conditions (12.12) and (12.13), the direction vectors $\mathbf{d}$ and $(\mathbf{h}_\pi, \mathbf{h}_v)$ are as follows:

$$\begin{aligned}\mathbf{d} &= -\mathbf{Z}\left(\mathbf{Z}^T\mathbf{D}_X^{-2}\mathbf{Z}\right)^{-1}\mathbf{Z}^T\nabla_x\Phi_\rho\left(\mathbf{x}^{(k)}, \pi^{(k)}, \mathbf{v}^{(k)}\right)\\ &= -\mathbf{D}_X\mathbf{P}\mathbf{D}_X\nabla_x\Phi_\rho\left(\mathbf{x}^{(k)}, \pi^{(k)}, \mathbf{v}^{(k)}\right), \qquad (12.15)\end{aligned}$$

where $\mathbf{Z}$ is a matrix of full rank that spans the null space of $\mathbf{A}$, the projection matrix $\mathbf{P} = \mathbf{I} - \mathbf{D}_X\mathbf{A}^T\left(\mathbf{A}\mathbf{D}_X^2\mathbf{A}^T\right)^{-1}\mathbf{A}\mathbf{D}_X$, and

$$\begin{bmatrix}\mathbf{h}_\pi\\ \mathbf{h}_v\end{bmatrix} = -\begin{bmatrix}\mathbf{I}\\ -\mathbf{A}^T\end{bmatrix}\left(\mathbf{A}\mathbf{D}_V^{-2}\mathbf{A}^T\right)^{-1}\mathbf{A}\nabla_v\Phi_\rho\left(\mathbf{x}^{(k)}, \pi^{(k)}, \mathbf{v}^{(k)}\right). \qquad (12.16)$$

In the last expression, $\left(\mathbf{A}\mathbf{D}_V^{-2}\mathbf{A}^T\right)^{-1} = \left(\mathbf{A}\mathbf{D}_X^2\mathbf{A}^T\right)^{-1}$. Thus only a single factorization is needed in order to compute the primal and dual directions by successive back substitutions.

Proposition 12.1: The directions $\mathbf{d}$ and $(\mathbf{h}_\pi, \mathbf{h}_v)$ are directions of descent for the components of the gradient vector of Φ_ρ in the primal and dual spaces, respectively.

This result follows directly from the above expressions for the directions or from their derivation based on (12.9) and (12.10), where, in the case of the dual, convert (12.10) into a minimization problem by reversing the sign of the objective.

The potential function Φ_ρ approaches $+\infty$ as the boundaries of the feasible primal and dual polytopes are approached. Thus, a *conceptual* algorithm can be defined by choosing the next iterate to be the minimizing point along the above primal and dual search directions from the current iterate. An *implementable* algorithm can be developed by using Wolfe-type exit conditions on the line search. For example, we can use the line search of Section 5.4, modified to impose an upper bound on the step length. Let us denote this next iterate by $\left(\mathbf{x}^{(k+1)}, \pi^{(k+1)}, \mathbf{v}^{(k+1)}\right)$ and the common step length along the search direction to this point by $t^{(k)}$.

The complexity analysis of the algorithm is based on the following two results:

Proposition 12.2: The directional derivative of the potential function along $(\mathbf{d}, \mathbf{h}_\pi, \mathbf{h}_v)$ is bounded above by $-\frac{3}{4}\max_i \left(x_i^{(k)} v_i^{(k)}\right)^2$.

Proposition 12.3: When $\rho \geq n + \sqrt{n}$, there exists a positive step length $t^{(k)}$ along the search direction such that

$$\Phi_\rho\left(\mathbf{x}^{(k+1)}, \pi^{(k+1)}, \mathbf{v}^{(k+1)}\right) - \Phi_\rho\left(\mathbf{x}^{(k)}, \pi^{(k)}, \mathbf{v}^{(k)}\right) \geq \tfrac{1}{8}.$$

Proofs of these two propositions are quite straightforward, and they imply, in turn, that the algorithm has polynomial-time complexity. Details can be found in the survey article of Todd [1996, Sec. 6].

12.3 Discussion

The Lagrangian optimality conditions (12.12), or equivalently (12.13), can be reformulated as follows, using (12.14) and the definition $\left(\vec{\mathbf{d}}, \vec{\lambda}, \vec{\mathbf{z}}\right) = (\mu\mathbf{d}, \mu\lambda, \mu\mathbf{z})$, where $\mu = \sigma\mu^{(k)}$ is given by (12.5)–(12.7),

$$\begin{aligned} \mathbf{A}\vec{\mathbf{d}} &= \mathbf{0}, \\ \mathbf{A}^T\vec{\lambda} + \vec{\mathbf{z}} &= \mathbf{0}, \\ \mathbf{V}_k\vec{\mathbf{d}} + \mathbf{X}_k\vec{\mathbf{z}} &= -\mathbf{X}_k\mathbf{V}_k\mathbf{e} + \mu\mathbf{e}. \end{aligned} \tag{12.17}$$

Note that the foregoing equations are precisely the equations defining the Newton corrector direction of Section 10.2.4 for the homotopy system (10.2), when the latter are specialized to the central path case and feasible iterates. The Newton corrector directions are the same as the directions of the primal–dual potential-reduction algorithm up to a scalar multiple μ. And the components of the Newton direction at the current iterate are *descent* directions for the potential function as shown by Proposition 12.1 of the previous section. *Thus there is an intimate connection between potential-reduction and path-following.*

The primal–dual potential-reduction algorithm of the previous section is akin to the techniques described in Section 10.5 when a *nonadaptive* target point is used—see the discussion immediately after expression (12.7)—and the second-order corrector is omitted. It is interesting to compare the directions generated by these two approaches at a (conceptual) iterate that actually lies *on the central path.* If the pure Euler or primal–dual affine-scaling direction proves to be effective, then the adaptively chosen centering parameter $\sigma^{(k)}$ in Mehrotra's algorithm is set to a value close to zero. Under these circumstances, the direction used in the Mehrotra algorithm is essentially the pure Euler direction with a second-order correction. In the potential-reduction algorithm, on the other hand, the quantity $\rho \equiv n/\sigma$ is fixed throughout; i.e., a *fixed* target is used and the direction is a Newton corrector to a predetermined target point. The Mehrotra approach can diminish (or enhance) centering adaptively, whereas the potential-reduction approach is less flexible.

From its form of derivation, using the Newton–Cauchy perspective, we see that the potential-reduction algorithm also has characteristics akin to the primal–dual affine scaling algorithm (Chapter 8). Like an affine-scaling algorithm, the potential-reduction algorithm is an *inherently* polytope-interior,[2] long-step approach. In contrast to path-following based on neighborhoods and targets, or primal–dual affine scaling based on direct reduction of the duality gap, the potential-reduction algorithm uses a line search applied to the primal–dual potential function. *It can be viewed as occupying the middle ground between affine scaling and Euler–Newton path-following.*

The formulation of the potential-reduction algorithm, paralleling primal–dual affine scaling, highlights the opportunities for enhancing efficiency using the null-space forms of the search directions, iterative methods for solving equations, and inexact computation, along lines very similar to those discussed in Chapter 7; see also Kim and Nazareth [1994]. With affine scaling, primal and dual directions and steps along them are decoupled. The potential function couples them together and takes a common step in primal–dual space. This suggests development of a hybrid of primal–dual potential reduction and affine scaling akin to the approach underlying Mehrotra's implementation, where the quantity ρ in the potential function is chosen adaptively and different steps are taken in the primal and dual spaces. Clearly, there are ample opportunities for novel and more effective implementation of a potential-reduction approach.

Other generalizations can be envisioned based on exploiting weighted potentials and their relationship to homotopy paths as discussed in Section 12.1.3. Details are left to the reader.

Infeasible starting points for the potential-reduction algorithm can be implemented by adding an artificial column/row to the LP matrix, as is commonly done in implementing affine-scaling algorithms. Infeasibility extensions at a more fundamental level can be motivated by (11.10) when the point $\left(\mathbf{x}^{(0)}, \pi^{(0)}, \mathbf{v}^{(0)}\right)$ is *not* a feasible interior point for (P) and (D) and the associated vectors $\mathbf{q}_{x^{(0)}}$ and $\mathbf{q}_{v^{(0)}}$ are nonzero. A development paralleling that pursued earlier for the feasible case in Sections 12.1 and 12.2 can be envisioned. However, our view is that the potential-reduction approach is *inherently* polytope-interior as opposed to orthant-interior—see Section 10.6 for a definition of these terms—and that root-level infeasibility extensions of this type may not be worth the effort.

---

[2]For terminology, see Section 10.6.

12.4 Concluding Remarks

Although the specifics of Karmarkar's polynomial-time algorithm—a nonstandard form of the linear program, projective transformations, and a primal potential function—have all been supplanted, his algorithm contained the seeds for many subsequent interior point developments, for example, the simpler (rediscovered) affine-scaling, the central path, and the primal–dual potential function, and it served to revitalize log-barrier methods and Euler–Newton path-following. Thus, one can only marvel at the extraordinary prescience in this landmark contribution and at the enormous literature it engendered; see Kranich [1991] for a bibliography of over 2000 research articles—that, in turn, transformed the post-1984 landscape of differentiable programming.

12.5 Notes

Section 12.0: Projective transformations used by Karmarkar [1984] have been displaced by the simpler (rediscovered) affine-scaling transformations of Dikin [1967]. Although Karmarkar's projective transformations are no longer viewed as being central to algorithmic linear programming, they remain interesting from a mathematical standpoint, and in particular, connections have been established to the collinear-scaling transformations and conic algorithms of Davidon [1980]; see Gonzaga [1989], Lagarias [1993], and Nazareth [1993b], [1994c], [1995d] for some early developments along these lines.

For example, conic models can be used to develop Karmarkar-like algorithms for linear programming, wherein the Dikin ellipsoidal trust region of Section 8.1.1 is replaced by a trust region defined as follows:

$$\frac{\left(\mathbf{x}-\mathbf{x}^{(k)}\right)^T \mathbf{X}_k^{-2}\left(\mathbf{x}-\mathbf{x}^{(k)}\right)}{\left(1+\mathbf{h}^T \mathbf{X}_k^{-1}\left(\mathbf{x}-\mathbf{x}^{(k)}\right)\right)^2} \leq \rho^2, \tag{12.18}$$

where $\mathbf{h} = \gamma\mathbf{e}$, $\gamma \geq 0$ is a so-called gauge parameter, and $\mathbf{e}$ is a vector of 1's. The restriction $\rho < 1/||\mathbf{h}||$ ensures that the foregoing conic-based trust region is ellipsoidal. The trust region (12.18) lies in the nonnegative orthant, and by choosing γ suitably, one can arrange that it touches the coordinate hyperplanes as in Dikin's ellipsoid. It is fitting to call this trust region the *Davidon ellipsoid.* For further discussion, see Nazareth [1995d]. Many interesting issues remain to be explored in this area, and similar approaches can also be formulated in dual and primal–dual settings.

Section 12.1: The primal–dual potential function described in this section is due to Tanabe [1988] and Todd and Ye [1990].

Section 12.2: The algorithm described in this section is the primal–dual potential-reduction algorithm of Kojima, Mizuno, and Yoshise [1991]. The formulation of its search direction is due to Todd [1996], which is modeled on a similar development for the primal–dual affine-scaling algorithm given in Nazareth [1994c], [1998]; see Chapter 8.

Part V

Algorithmic Science

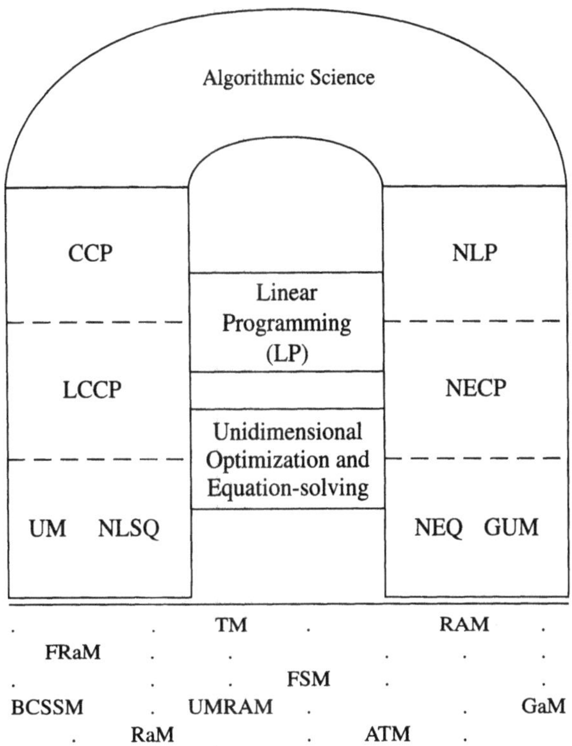

13

Algorithmic Principles

We now focus on a specific optimization method—the variable-metric method for unconstrained minimization—and consider a set of underlying algorithmic principles that have served to guide and shape its development. The discussion highlights the very elegant structure of interrelationships between variable-metric formulae that has emerged during the course of their investigation. It also seeks justification for the grassroots belief that the algorithm based on the BFGS update is the best member of the variable-metric family.

In the concluding section of this chapter we reflect on the "essentialist" philosophy of algorithmic development that underlies the oft-pursued task of finding the "best" member of a family of algorithms, as exemplified by the variable-metric *case study* of the present chapter. This discussion provides motivation for an alternative, "nonessentialist," view of algorithms that is presented in the next chapter.

13.1 Introduction

The BFGS variable-metric algorithm was introduced in Chapter 2. Its conceptual form uses an exact line search and proceeds as follows:

Algorithm BFGS

Given starting point $\mathbf{x}_1$, gradient $\mathbf{g}_1$, and a positive definite initial Hessian approximation, typically, $\mathbf{M}_1 = \mathbf{I}$:

for $k = 1, 2, \ldots$

Solve $\mathbf{M}_k \mathbf{d}_k = -\mathbf{g}_k$ for $\mathbf{d}_k$.

$\alpha_k^* = \text{argmin}_{\alpha \geq 0} f(\mathbf{x}_k + \alpha \mathbf{d}_k)$.

$\mathbf{x}_{k+1} = \mathbf{x}_k + \alpha_k^* \mathbf{d}_k$. Evaluate $\mathbf{g}_{k+1}$.

$\mathbf{s}_k = \mathbf{x}_{k+1} - \mathbf{x}_k$; $\mathbf{y}_k = \mathbf{g}_{k+1} - \mathbf{g}_k$; $a_k = \mathbf{s}_k^T \mathbf{M}_k \mathbf{s}_k$; $b_k = \mathbf{y}_k^T \mathbf{s}_k$;

$\mathbf{M}_{k+1} = \mathbf{M}_k - \frac{\mathbf{M}_k \mathbf{s}_k \mathbf{s}_k^T \mathbf{M}_k}{a_k} + \frac{\mathbf{y}_k \mathbf{y}_k^T}{b_k}$.

end

The algorithm uses the BFGS update defined by (2.12). When the starting matrix $\mathbf{M}_1$ is positive definite, all subsequent matrices $\mathbf{M}_k$, $k \geq 2$, will remain positive definite.[1] In an implementable version of the algorithm, the exact line search is replaced by a line search based on the Wolfe exit conditions as discussed in Chapter 5, and again it is easy to show that the updated matrices $\mathbf{M}_k$, $k \geq 2$, remain positive definite. For details, see, for example, Nazareth [1994b].

The original version of the variable-metric algorithm, which was discovered by Davidon [1959] and refined by Fletcher and Powell [1963], was the same, in essence, as the above algorithm. But it used a different updating formula, named the DFP (after its pioneers). As with Karmarkar's algorithm two decades later, Davidon's breakthrough generated a huge outpouring of new research results on variable-metric updates, over an extended period of time. During the course of this research, the BFGS algorithm was shown to possess very attractive properties: finite termination when applied to a strictly convex quadratic function, global convergence on a strictly convex function, superlinear convergence in the vicinity of a strict local minimum, global convergence on smooth functions when the condition number of the Hessian approximation matrix is explicitly controlled so that it does not exceed any prespecified, finite upper bound, and so on. For an overview, see Dennis and Schnabel [1983]. Many of these properties are shared with algorithms based on other variable-metric updates, but the BFGS has the strongest set of properties in totality, and it has emerged as the leading contender in theory and practice.

13.2 Duality

13.2.1 Complementing Operation

Suppose the positive definite matrix $\mathbf{M}_k$, the nonzero step $\mathbf{s}_k$, and its corresponding nonzero gradient change $\mathbf{y}_k$ are the quantities developed within

[1]The superscript +, which was used in Chapter 2 to explicitly distinguish positive definite matrices, is no longer needed here, and henceforth it will be omitted.

the BFGS algorithm of the previous section, and let $\mathbf{W}_k = \mathbf{M}_k^{-1}$. Define

$$a_k = \mathbf{s}_k^T \mathbf{M}_k \mathbf{s}_k, \qquad b_k = \mathbf{y}_k^T \mathbf{s}_k, \qquad c_k = \mathbf{y}_k^T \mathbf{W}_k \mathbf{y}_k.$$

Positive definiteness of the matrices $\mathbf{M}_k$ and $\mathbf{W}_k$ implies that the quantities a_k and b_k, which arise within the algorithm, are positive, and the use of an exact line search or the standard Wolfe-type exit conditions of Chapter 5 within an implementable version of a line search will ensure that b_k is also positive.

The BFGS update of $\mathbf{M}_k$, as seen in the algorithm of Section 13.1, is

$$\mathbf{M}_{k+1} = \mathbf{M}_k - \frac{\mathbf{M}_k \mathbf{s}_k \mathbf{s}_k^T \mathbf{M}_k}{a_k} + \frac{\mathbf{y}_k \mathbf{y}_k^T}{b_k}, \tag{13.1}$$

and it satisfies the quasi-Newton relation $\mathbf{M}_{k+1}\mathbf{s}_k = \mathbf{y}_k$.

Make the following *formal interchanges of symbols* in the BFGS update formula:

$$\mathbf{M} \leftrightarrow \mathbf{W}, \qquad \mathbf{s} \leftrightarrow \mathbf{y}, \tag{13.2}$$

and retain the subscripts. This symbolic interchange operation is called *dualizing* or *complementing* an update,[2] and when applied to the BFGS update, it yields the *complementary-BFGS update* formula:

$$\mathbf{W}_{k+1} = \mathbf{W}_k - \frac{\mathbf{W}_k \mathbf{y}_k \mathbf{y}_k^T \mathbf{W}_k}{c_k} + \frac{\mathbf{s}_k \mathbf{s}_k^T}{b_k}. \tag{13.3}$$

Since the BFGS update satisfies $\mathbf{M}_{k+1}\mathbf{s}_k = \mathbf{y}_k$, the complementing operation implies that the matrix $\mathbf{W}_{k+1}$ satisfies the quasi-Newton relation defined in terms of approximations to the *inverse* Hessian, namely, $\mathbf{W}_{k+1}\mathbf{y}_k = \mathbf{s}_k$.

Given the positive definite matrix $\mathbf{M}_k$ and its inverse $\mathbf{W}_k \equiv \mathbf{M}_k^{-1}$, we can pose the following question: Is the matrix obtained by using the complementary-BFGS formula (13.3) the *same* as the inverse of the matrix obtained using the BFGS formula (13.1)? For clarity, let us write the matrix obtained from the BFGS update of $\mathbf{M}_k$ as $\mathbf{M}_{k+1}^B$ and the matrix obtained from the complementary-BFGS update of $\mathbf{W}_k \equiv \mathbf{M}_k^{-1}$ as $\mathbf{W}_{k+1}^D$. Then, is the matrix $\mathbf{M}_{k+1}^B$ the same as the matrix $(\mathbf{W}_{k+1}^D)^{-1}$? The answer is *no*! The complementing operation yields a different update from the BFGS. In fact, the complementary-BFGS update is the original DFP update mentioned above.[3]

---

[2] "Dualizing" is the *generic* operation of symbolic interchange to obtain a related object, and this term is commonly used in the variable-metric literature. However, we prefer to use "complementing" an update to avoid confusion with the dualizing of a mathematical program.

[3] Historically, the derivation came about in reverse; i.e., the BFGS update was discovered by Fletcher [1970] by applying the complementing operation to the original DFP update.

The BFGS and DFP updates are called a *complementary pair* of updates. In order to explore the issue of complementary updates further, let us now consider a broader family.

13.2.2 The Broyden Family

There are infinitely many updates that satisfy the QN relation. Broyden [1970] defined a one-parameter family of rank-2 updates of $\mathbf{M}_k$ to $\mathbf{M}_{k+1}(\mu)$, which embraces most updates nowadays in common use, as follows:

$$\mathbf{M}_{k+1}(\mu) = \mathbf{M}_{k+1}(0) + (\mu a_k)\mathbf{m}_k\mathbf{m}_k^T, \tag{13.4}$$

where $\mu \in R$ is a parameter. The matrix $\mathbf{M}_{k+1}(0)$, which forms the root of the family, is the BFGS-update (13.1). It was previously denoted by $\mathbf{M}_{k+1}$ or, for clarity, $\mathbf{M}_{k+1}^B$, and it is now written instead with the parameter value $\mu = 0$ explicitly attached. The vector $\mathbf{m}_k$ is defined by

$$\mathbf{m}_k = \frac{\mathbf{y}_k}{b_k} - \frac{\mathbf{M}_k\mathbf{s}_k}{a_k}. \tag{13.5}$$

Note that $\mathbf{m}_k^T\mathbf{s}_k = 0$, and thus $\mathbf{M}_{k+1}(\mu)\mathbf{s}_k = \mathbf{y}_k \,\forall\, \mu$.

The complementary-Broyden family of updates of $\mathbf{W}_k$ to $\mathbf{W}_{k+1}(\nu)$, depending on a parameter $\nu \in R$, is as follows:

$$\mathbf{W}_{k+1}(\nu) = \mathbf{W}_{k+1}(0) + (\nu c_k)\mathbf{w}_k\mathbf{w}_k^T, \tag{13.6}$$

where $\mathbf{W}_{k+1}(0)$, the root of the family, is the complementary-BFGS or DFP update (13.3), now also written with the parameter value explicitly attached, and

$$\mathbf{w}_k = \frac{\mathbf{s}_k}{b_k} - \frac{\mathbf{W}_k\mathbf{y}_k}{c_k}. \tag{13.7}$$

The family is obtained by applying the complementing operation to (13.4). Note also that (13.2) implies that a_k is complementary to c_k, b_k is self-complementary, and $\mathbf{m}_k$ is complementary to $\mathbf{w}_k$. The quantity $\nu \in R$ defines the complementary parameterization.

Given $\mathbf{M}_k = \mathbf{W}_k^{-1}$, the following relationship between the parameters μ and ν ensures that $\mathbf{M}_{k+1}(\mu)$ and $\mathbf{W}_{k+1}(\nu)$ are inverses of one another:

$$\mu + \nu - \mu\nu(1 - ac/b^2) = 1. \tag{13.8}$$

This can be verified by directly multiplying the two matrices $\mathbf{M}_{k+1}(\mu)$ and $\mathbf{W}_{k+1}(\nu)$, with μ and ν related by (13.8). Note that (13.8) is satisfied by $\mu = 0$ and $\nu = 1$, and by $\nu = 0$ and $\mu = 1$. Thus the *inverse* of $\mathbf{M}_{k+1}(0)$ is $\mathbf{W}_{k+1}(1)$; i.e., the choice $\nu = 1$ in (13.6) gives the BFGS update of the inverse Hessian approximation $\mathbf{W}_k$. On the other hand, as noted earlier, the *complement* of $\mathbf{M}_{k+1}(0)$ is $\mathbf{W}_{k+1}(0)$, namely, the DFP update of $\mathbf{W}_k$.

13.2.3 Self-Complementary Updates

Are there updates in the Broyden family that yield the *same* update under the complementing operation? An example of such an update is the symmetric rank-1 (SR1) given by expression (2.6). It is obtained from the Broyden family by making the choice of parameters μ and ν as follows:

$$\mu_{\mathrm{SR1}} = \frac{b_k}{b_k - a_k}, \qquad \nu_{\mathrm{SR1}} = \frac{b_k}{b_k - c_k}. \tag{13.9}$$

Using this choice and also dropping explicit attachment of the parameter to the updated matrix, the Broyden update formula simplifies to

$$\mathbf{M}_{k+1} = \mathbf{M}_k + \frac{(\mathbf{y}_k - \mathbf{M}_k\mathbf{s}_k)(\mathbf{y}_k - \mathbf{M}_k\mathbf{s}_k)^T}{b_k - a_k}, \tag{13.10}$$

and its inverse $\mathbf{W}_{k+1} = \mathbf{M}_{k+1}^{-1}$, which can be derived, for example, using the well-known Sherman–Morrison formula, is as follows:

$$\mathbf{W}_{k+1} = \mathbf{W}_k + \frac{(\mathbf{s}_k - \mathbf{W}_k\mathbf{y}_k)(\mathbf{s}_k - \mathbf{W}_k\mathbf{y}_k)^T}{b_k - c_k}, \tag{13.11}$$

where $\mathbf{W}_k = \mathbf{M}_k^{-1}$. The identical expression is obtained by applying the complementing operation (13.2) to (13.10); i.e., *the operations of complementing and inverting produce the same update.* The SR1 update is said to be *self-complementary* or self-dual.

It can be shown that there are infinitely many self-complementary updates within the Broyden family; i.e., the family of update formulae contains very interesting mathematical substructure. For a detailed discussion, see Nazareth and Zhu [1995], Zhu [1997], and references cited therein. Unfortunately, none of the self-complementary updates studied to date have proved, in practice, to be significantly and consistently superior to the BFGS. Self-complementarity appears to be an elegant *algebraic property* that some updating formulae possess, a posteriori, rather than a fundamental *algorithmic principle* to which a successful update should conform, a priori.

13.3 Invariance

The BFGS update formula is invariant under a transformation of variables. This can be seen as follows. Let $\mathbf{x}_k$, $\mathbf{g}_k$, $\mathbf{M}_k$, $\mathbf{s}_k$, $\mathbf{y}_k$, $k = 1, 2, \ldots,$ denote the sequence of iterates, gradients, Hessian approximations, steps, and gradient changes that are generated by the BFGS algorithm of Section 13.1. Let $\mathbf{P}$ be an arbitrary nonsingular matrix. Make the *fixed* transformation of variables or preconditioning $\tilde{\mathbf{x}} = \mathbf{P}\mathbf{x}$. Then it is easily verified by the chain rule that the foregoing quantities transform as follows: The gradient $\mathbf{g}$

and (true) Hessian $\mathbf{H}$ at any point $\mathbf{x}$ transform to $\tilde{\mathbf{g}} = \mathbf{P}^{-T}\mathbf{g}$ and $\tilde{\mathbf{H}} = \mathbf{P}^{-T}\mathbf{H}\mathbf{P}^{-1}$; steps transform like points; i.e., $\tilde{\mathbf{s}}_k = \mathbf{P}\mathbf{s}_k$, and gradient changes like gradients; i.e., $\tilde{\mathbf{y}}_k = \mathbf{P}^{-T}\mathbf{y}_k$. Hessian approximations transform like the Hessian $\mathbf{H}$; i.e., $\widetilde{\mathbf{M}}_k = \mathbf{P}^{-T}\mathbf{M}_k\mathbf{P}^{-1}$.

Consider the BFGS update formula in the *transformed* space; i.e.,

$$\widetilde{\mathbf{M}}_{k+1} = \widetilde{\mathbf{M}}_k - \frac{(\widetilde{\mathbf{M}}_k\tilde{\mathbf{s}}_k)(\widetilde{\mathbf{M}}_k\tilde{\mathbf{s}}_k)^T}{\tilde{\mathbf{s}}_k^T\widetilde{\mathbf{M}}_k\tilde{\mathbf{s}}_k} + \frac{\tilde{\mathbf{y}}_k\tilde{\mathbf{y}}_k^T}{\tilde{\mathbf{y}}_k^T\tilde{\mathbf{s}}_k}.$$

Directly substituting for these quantities yields

$$\mathbf{P}^{-T}\mathbf{M}_{k+1}\mathbf{P}^{-1} = \mathbf{P}^{-T}\left[\mathbf{M}_k - \frac{(\mathbf{M}_k\mathbf{P}^{-1}\mathbf{P}\mathbf{s}_k)(\mathbf{M}_k\mathbf{P}^{-1}\mathbf{P}\mathbf{s}_k)^T}{\mathbf{s}_k^T\mathbf{P}^T\mathbf{P}^{-T}\mathbf{M}_k\mathbf{P}^{-1}\mathbf{P}\mathbf{s}_k} + \frac{\mathbf{y}_k\mathbf{y}_k^T}{\mathbf{y}_k^T\mathbf{P}^{-1}\mathbf{P}\mathbf{s}_k}\right]\mathbf{P}^{-1}.$$

And simplifying this expression gives

$$\mathbf{M}_{k+1} = \mathbf{M}_k - \frac{(\mathbf{M}_k\mathbf{s}_k)(\mathbf{M}_k\mathbf{s}_k)^T}{\mathbf{s}_k^T\mathbf{M}_k\mathbf{s}_k} + \frac{\mathbf{y}_k\mathbf{y}_k^T}{\mathbf{y}_k^T\mathbf{s}_k}.$$

We see immediately that the *updating formula is invariant* under the transformation of variables defined by $\mathbf{P}$. Indeed, *any member of the Broyden family (13.4) shares this invariance property,* which can be verified in an analogous way. Only the initialization changes; i.e., $\widetilde{\mathbf{M}}_1 = \mathbf{P}^{-T}\mathbf{M}_1\mathbf{P}^{-1}$.

Subject to an appropriate choice of the initialization, the BFGS algorithm of Section 13.1, or an analogous variable-metric algorithm based on any other member of the Broyden family, can be seen to be invariant under a transformation of variables $\tilde{\mathbf{x}} = \mathbf{P}\mathbf{x}$. In contrast, the formulae for defining CG-related search directions discussed in Chapter 5 are *not* invariant under a transformation of variables; i.e., in the transformed space, the matrix $\mathbf{P}$ enters explicitly into the defining expressions. The nonlinear CG algorithm of Section 5.2.1 is not invariant under a transformation of variables.

13.4 Symmetry

Consider any sequence of linearly independent steps $\mathbf{s}_1, \mathbf{s}_2, \ldots \mathbf{s}_k$, $k \leq n$, with associated gradient changes $\mathbf{y}_1, \mathbf{y}_2, \ldots, \mathbf{y}_k$. Starting with an initial positive definite matrix, say $\mathbf{M}_1$, suppose a variable-metric update is performed using these steps in succession, yielding the matrix $\mathbf{M}_{k+1}$. *Does the matrix* $\mathbf{M}_{k+1}$ *depend upon the order in which the steps are used?*

For an *arbitrary* differentiable function, say f, the answer to this question is obviously yes, because in general, the quasi-Newton relation will be satisfied by $\mathbf{M}_{k+1}$ only over the *most recently used* step. This is true for almost all updates in the Broyden family, even when f is further restricted to be a strictly convex quadratic function.[4] Therefore, let us examine the foregoing question within the more specialized setting of a strictly convex quadratic function and a set of k steps $\mathbf{s}_1, \mathbf{s}_2, \ldots, \mathbf{s}_k$ that are *mutually conjugate* with respect to the positive definite Hessian matrix $\mathbf{A}$ of the quadratic.

Define the following procedure that applies the BFGS update—for convenience used in its inverse form (2.13)—to the given pairs of steps and corresponding gradient changes, taken in sequence.

Procedure S

Given $\mathbf{W}_1$ positive definite, symmetric and $k \leq n$ mutually *conjugate* steps $\mathbf{s}_j \neq 0$ with corresponding gradient changes $\mathbf{y}_j$, $1 \leq j \leq k$:

for $j = 1, \ldots, k$

$$\mathbf{W}_{j+1} = \left(\mathbf{I} - \frac{\mathbf{s}_j \mathbf{y}_j^T}{\mathbf{y}_j^T \mathbf{s}_j}\right) \mathbf{W}_j \left(\mathbf{I} - \frac{\mathbf{s}_j \mathbf{y}_j^T}{\mathbf{y}_j^T \mathbf{s}_j}\right)^T + \frac{\mathbf{s}_j \mathbf{s}_j^T}{\mathbf{y}_j^T \mathbf{s}_j} \tag{13.12}$$

end

When $k = n$, it is well known that $\mathbf{W}_{n+1} = \mathbf{A}^{-1}$; i.e., Procedure S is a way to infer the inverse of a positive definite symmetric matrix $\mathbf{A}$ via variable-metric updating. When $k \leq n$, the following proposition, which concerns the matrices produced by Procedure S, is easily proved by induction:

Proposition 13.1: At each stage $j \leq k$ of Procedure S, the matrix $\mathbf{W}_{j+1}$ satisfies the following condition:

$$\mathbf{W}_{j+1} = \left(\mathbf{I} - \sum_{i=1}^{j} \frac{\mathbf{s}_i \mathbf{y}_i^T}{\mathbf{y}_i^T \mathbf{s}_i}\right) \mathbf{W}_1 \left(\mathbf{I} - \sum_{i=1}^{j} \frac{\mathbf{s}_i \mathbf{y}_i^T}{\mathbf{y}_i^T \mathbf{s}_i}\right)^T + \sum_{i=1}^{j} \frac{\mathbf{s}_i \mathbf{s}_i^T}{\mathbf{y}_i^T \mathbf{s}_i}. \tag{13.13}$$

Proof: See Nazareth [1994b, pp. 20–21]. □

Expression (13.13) shows that the matrix $\mathbf{W}_{k+1}$ is invariant with respect to permutations of the steps $\mathbf{s}_1, \ldots, \mathbf{s}_k$. Thus, if the steps $\mathbf{s}_1, \ldots, \mathbf{s}_k$ had been used in a *different order* in *Procedure S*, the *same* matrix $\mathbf{W}_{k+1}$ would have resulted after k successive updates. This result is true for $k = n$, as noted above, and we now see that it also holds for all intermediate values of k.

---

[4]An exception is the SR1 update. When this update is applied to a strictly convex quadratic function over *any* k linearly independent steps, and it does not break down at any stage because of a zero denominator, then the quasi-Newton relation is satisfied over *all* k steps, not just the most recent one.

It is of particular relevance to limited-memory BFGS updating algorithms described in Chapter 2.

Complementarity between the BFGS and DFP updates described in Section 13.2 implies that Proposition 13.1 and its conclusion concerning update symmetry also hold for the DFP update. It would be interesting to explore their validity for other updates in the Broyden family, for example, the SR1.

13.5 Conservation

13.5.1 Variational Principles

Conservation, in the present context, involves the preservation of information in the matrix approximating the Hessian between successive iterations of a variable-metric algorithm, through the use of appropriate variational principles.

The derivation of the BFGS update in Section 2.3.2 is based on conservation principles for the reconditioner $\mathbf{R}_k$. First, $\mathbf{R}_k$ is permitted to be modified only by addition of a matrix of rank 1. This effects a minimal change, because $\mathbf{R}_{k+1}$ and $\mathbf{R}_k$ leave an $(n-1)$-dimensional subspace invariant. Secondly, the quantity $\|\mathbf{R}_{k+1} - \mathbf{R}_k\|$, where $\|\cdot\|$ denotes the Frobenius or spectral matrix norm, is minimized subject to $\mathbf{R}_{k+1}$ satisfying the quasi-Newton relation in the form (2.9).

Suppose $\mathbf{R}_k^{-1}$ is substituted for $\mathbf{R}_k$ in the foregoing derivation. Then the discussion on complementary updates of Section 13.2 implies that the DFP update would be obtained from the corresponding variational (or conservation) principle.

Conservation principles along similar lines, which involve minimization of the deviation, in various norms, between the Hessian approximations $\mathbf{M}_{k+1}$ and $\mathbf{M}_k$, or between their inverses $\mathbf{W}_{k+1}$ and $\mathbf{W}_k$, have been studied in great detail in the literature. See, in particular, Dennis and Moré [1977] and Dennis and Schnabel [1983].

In an alternative approach, the deviation from the quasi-Newton *relation* on the step immediately prior to the most recent step $\mathbf{s}_k$ is minimized in an appropriate vector norm; see Nazareth [1984]. This yields the least prior deviation (LPD) update, which can be shown, a posteriori, to possess the self-complementarity property (Mifflin and Nazareth [1994]). The LPD update is *algebraically* distinct from the BFGS, but when line searches are exact, the LPD update can be reexpressed and reduces to the BFGS update, and thus *conceptual* algorithms based on the two updating formulae produce *numerically identical* Hessian-approximation matrices at corresponding iterates. For a computational study of the LPD update, see Zhu [1997].

13.5.2 Continuity

Conservation can also be used in the sense of continuity or preservation of an algorithmic formula across the borderline between two different families of algorithms. The relationship between the BFGS variable-metric algorithm and the nonlinear conjugate gradient algorithm, which was discussed in Section 2.3.3, provides an example. The continuity between the LPD and BFGS updates under exact line searches, as just noted at the end of Section 13.5.1, provides another example.

To further highlight the *structural correspondence* between the BFGS and CG algorithmic formulae, consider the behavior of any member of the Broyden family of updates (13.4) *when applied to a strictly convex quadratic function.* Thus, suppose the update of $\mathbf{M}_k$ to $\mathbf{M}_{k+1}$ in Algorithm BFGS of Section 13.1 is replaced by any update in (13.4), where μ is restricted to values $\mu \geq \mu^*$ that preserve positive definiteness of the Hessian approximation.[5] Assume $\mathbf{M}_1 = \mathbf{I}$. Let $\mathbf{d}_k^\mu$ denote the search direction, at iteration k, developed by this Broyden procedure. Let $\mathbf{d}_k^{CG}$ denote the corresponding search direction developed by the CG procedure of Sections 5.2.1 and 5.5.1, using any of the equivalent definitions of the quantity β_k, for example, the Fletcher–Reeves choice. It is well known that the direction $\mathbf{d}_k^\mu$ is *parallel* to the corresponding direction $\mathbf{d}_k^{CG}$, *for any choice of the parameter* μ. Now, for one particular member of the Broyden family—the BFGS update—this condition can be strengthened as follows:

$$\mathbf{d}_k^B = \mathbf{d}_k^{CG}, \quad 1 \leq k \leq n, \tag{13.14}$$

where $\mathbf{d}_k^B$ denotes the search direction in Algorithm BFGS at iteration k. The proof, by induction, is straightforward; see Nazareth [1976], [1994b].

A detailed discussion of the BFGS-CG relationship is given in Nazareth [1976], [1979], Buckley [1978], and Kolda, O'Leary, and Nazareth [1998]. It provides the foundation for a wide variety of CG-related limited-memory algorithms, as outlined in Section 2.3.3.

13.6 The "Essentialist" View

In a thought-provoking and controversial essay, the eminent numerical analyst R. Hamming [1977] makes the following concluding observations:[6]

> I find that some algorithms are indeed either invariant under what appears to be the most appropriate class of transformation, or else can be made so by small changes. But I also find

---

[5]For details on the definition of μ^*, see, for example, Nazareth [1994b, p. 64].

[6]Italics in this quotation are ours.

> that in some areas there are at present no such algorithms. How can I rely on them?
>
> Experience using this criterion shows that it can be very helpful in creating an algorithm. As one would expect, the larger the class of transformations the more restricted is the algorithm, and often the easier to find.
>
> This is how I was led to preaching the heresy that the test of the appropriateness of formulas (or all of the mathematics one uses) does not come from mathematics, but from the *physics of the situation*, and from the relevant broad principles of *invariance, conservation, and symmetry*. It would appear at first glance that I have merely replaced one set of arbitrary assumptions (mathematics) with another. But I feel that there is a profound difference, the basis of science[7] seems to me to be far more reliable than does much of modern mathematics.

In the spirit of the foregoing remarks, we have presented a case study for a particular family of algorithms based on the fundamental variable-metric/low-rank updating idea of Davidon[8] [1959]. Subsidiary principles, such as the ones discussed in previous sections of this chapter, which are often derived from physical science and mathematics and given *an algorithmic interpretation*, can then be brought to bear, in order to help identify the "best" algorithm within the family. Individual variable-metric algorithms, viewed as *mathematical objects* of both intrinsic interest and practical value, are seen to have fascinating properties, and they can be subjected to a variety of further analyses: rate of convergence, complexity, numerical, performance on a test bed of artificial and realistic problems. The BFGS-DFP complementary pair has emerged as a leading candidate, with the BFGS update being preferred, in general, over the DFP; see Powell [1986].

Other methods and derived families of optimization and equation-solving algorithms can be studied along analogous lines, using appropriate algorithmic principles. In earlier chapters of this monograph (Parts I–IV), a variety of such methods have been formulated, each premised on a key underlying algorithmic idea:

- Homotopy/Euler–Newton path-following in Section 3.2 and Chapter 10.
- Symmetry principle/Lagrangian potentials in Section 3.3.
- Conjugacy in Chapter 5.
- Continuity in Chapter 6.
- Affine scaling in Chapters 7 and 8.

---

[7]R. Hamming would probably have felt very comfortable with the label "algorithmic scientist."

[8]Coincidentally, a mathematical physicist by training.

- QC-relation/diagonal updating in Chapter 9.
- Symmetry principle/log-barrier potentials in Chapter 11.
- Karmarkar-type potentials/affine scaling in Chapter 12.

The foregoing foundational ideas lead to broad families of algorithms within which considerable effort has been devoted, by many researchers, to finding the most effective members. This approach to the study of algorithms is premised on an "essentialist" philosophy that a "best" algorithm exists within a family and can be discovered. To date, it has dominated the study of algorithms.

In the next chapter we present an alternative, nonessentialist, view.

13.7 Notes

Sections 13.1–13.5: This material is derived from Nazareth [1994b].

14
Multialgorithms: A New Paradigm

The previous chapter presented the "essentialist" view of algorithms for optimization and equation-solving. We now turn to an alternative, nonessentialist, perspective that is rooted in Darwinian, population-based ideas of evolutionary biology. A new *multialgorithms paradigm* is presented within the setting of a parallel-computing environment well suited to its implementation.

The focus in this chapter will be on multialgorithms for the unconstrained minimization problem. A detailed case study of CG-multialgorithms will be described along with computational experiments on a simulated multiple-instruction multiple-data (MIMD) machine.

Broader implications for optimization and equation-solving algorithms in general are discussed in the two concluding sections.

14.1 Background

14.1.1 Traditional UM Algorithms

Traditional unconstrained minimization algorithms discussed, in particular, in Chapters 2, 5, and 9 seek a *local* minimizing point of a nonlinear function f, over a domain of real-valued variables $x_1, \ldots, x_n$. The function f is typically smooth with one or more continuous derivatives, but it is not necessarily convex. It may have distinguishing topographical features, for example, kinks, flat regions, steeply curving valleys, that present significant challenges to an algorithm. And it may possess several local minima—

typically this number is not large—with no discernible correlation between individual minima. (Reinitiating the algorithm from a representative set of starting points is the standard way to discover an acceptable solution among them.) For purposes of discussion, we will subsequently refer to a function with the foregoing characteristics as a *type-1* function.

Recall the *two common assumptions* under which unconstrained optimization algorithms operate, as itemized in Section 2.1.1, namely:

1. The cost of obtaining information about the (type-1) objective function f is the dominant cost.
2. The routine that provides information about the objective function is difficult to parallelize.

Most techniques in current use for minimizing f are of Newton–Cauchy type; see Chapters 2 and 5. Let us revisit them briefly:

- *Newton's method* uses the Hessian matrix of f. Because this matrix can be indefinite, the natural way to formulate a Newton algorithm is to confine a quadratic approximation of the objective function f to a *trust region* defined by an ellipsoidal constraint centered on the current iterate. The resulting model or trust region subproblem is then used to obtain the next iterate; see Section 2.2.1. The focus shifts to identifying key quantities that define the shape and size of the trust region, the accuracy to which the solution of the subproblem is approximated, and the way the trust region is revised at each iteration. Much effort has been devoted to making good choices for these parameters based on convergence analysis and numerical experience. But what is "best" is difficult to say, if not impossible, and such choices are often made in an ad hoc manner.
- The *Variable-metric/quasi-Newton method* has generated an enormous literature following its discovery by Davidon [1959]. Most variable-metric updates in current use for developing approximations to the Hessian matrix belong to the one-parameter family of Broyden [1970]; see Section 13.2.2. A consensus has emerged that the BFGS update is the most effective member of this family. (There have been many partial explanations for this choice as outlined in Chapter 13.) However, it has also been noted that other members of the Broyden family can considerably outperform the BFGS on given problems. In addition to the Broyden parameter, other critical quantities within a variable-metric algorithm are as follows: the accuracy of the line search procedure; the initial step along the direction of search when the current iterate is not in the vicinity of a solution (it should approach a direct-prediction step near a solution); the Oren–Luenberger parameter for sizing the Hessian approximation; the length of the increment used

in a finite-difference procedure when gradients are estimated by finite differences.

- *The Conjugate-gradient method*, which was discovered by Hestenes and Stiefel [1952] and subsequently adapted to nonlinear minimization by Fletcher and Reeves [1964], also generated a very large literature on optimization techniques in the presence of limited available memory. A consensus has emerged that the Polyak–Polak–Ribière (PPR) variant is the most practical nonlinear CG algorithm, in spite of certain theoretical shortcomings; see Powell [1984]. In Chapter 5 we introduced a new two-parameter family of CG algorithms that subsumes most CG algorithms in current use. An "essentialist" avenue of research would be to seek the "best" CG algorithm within this family; i.e., the best choice of parameters in (5.10).
- *Cauchy's method* of steepest descent, with appropriate diagonal preconditioning and a suitable step-length strategy, is a good choice when the assumptions underlying the previous methods are violated. An excellent background discussion is given in Bertsekas [1999]. Stochastic quasi-gradient extensions of Cauchy's method have proved especially useful; see Ermoliev and Gaivoronski [1994] and articles in Ermoliev and Wets [1988].

A very elegant mathematical substructure has emerged that relates the "best" Newton–Cauchy methods to one another under certain ideal circumstances, for instance, when line searches are exact or search directions are conjugate. See the discussion in Chapters 2 and 13.

14.1.2 Evolutionary GUM Algorithms

The term evolutionary algorithms refers to a collection of closely interrelated techniques—genetic algorithms, evolution strategies, and evolutionary programming—that take their algorithmic metaphors from the field of evolutionary biology. In recent years there has been a maturing and convergence of these three basic approaches; see the monographs of Bäck [1996], D.B. Fogel [1995], Michalewicz [1996], Schwefel [1995], and the research literature cited therein, which are premised on the pioneering work of Bremermann, Holland, L.J. Fogel, Rechenberg, Schwefel, and others. They continue to be viewed with some suspicion by many researchers in traditional areas of computational optimization, mainly because they are perceived as heuristic and not amenable to rigorous mathematical analysis. For example, Fox [1995] declares current genetic algorithms to be "handicapped by slavish mimicking" of their biological origins, and seeks to reposition them as algorithms based on Markovian chains on a sufficiently rich state space, thereby also placing them under a common umbrella shared by simulated annealing and other probabilistic search algorithms that are based on physical analogies.

Evolutionary algorithms are useful for finding a good approximation to the *global* minimum of a function that has very different characteristics from the type-1 functions considered in the previous subsection. Examples of such functions can be found in Michalewicz [1996, Chapter 2] and Bäck [1996, Chapter 3]. Typically, such functions have *a very large number of local minima, often, though not necessarily, with significant correlations between them.* Henceforth, for convenience of discussion, we refer to them as (*correlated* or *uncorrelated*) *type-2* functions and denote them by the symbol F over a domain of real-valued parameters $\lambda_1, \ldots, \lambda_m$. (We intentionally use a notation different from that used for type-1 functions.) For a specific instance and a visual plot when $m = 2$; see Ackley's function in Bäck [1996, Section 3.3, p. 142]. Type-2 functions need not be continuous or differentiable, and indeed, they are commonly defined only over a grid or other discrete set of points in the domain.

Newton–Cauchy algorithms, metaphorically refered to as "a marble rolling downhill," are essentially useless when applied to a type-2 function. They rapidly get stuck at a local minimum that may be very far from global optimality. In contrast, evolutionary algorithms employ a different metaphor. As summarized by Bäck [1996, p. 35]:

> The idea is to use a simulated evolutionary process for the purpose of solving an optimization problem, where the goal is to find a set of parameters (which may be interpreted as a "genotype" as well as a "phenotype") such that a certain quality criterion is maximized or minimized.

And quoting Fogel [1992, Section 1], evolutionary algorithms

> operate on a population of candidate solutions, subject these solutions to alterations, and employ a selection criterion to determine which solutions to maintain for future generations.

The genetic metaphor is derived from the interpretation of F as a fitness function or *rugged fitness landscape*, the use of a *population* of which each member is a candidate solution of the problem (called a *genotype*), and the mimicking of the genetic operations of *recombination* and *mutation.* There is little distinction made between the notions of genotype and phenotype as used in this setting; see the above quotation from Bäck [1996] and especially Tables 2.6 and 2.5 (pp. 132–134) of this unifying monograph.

Kauffman [1993, p. 95], addressing a maximization problem, makes the following important observation:

> Recombination is useless on uncorrelated landscapes, but can be effective under two conditions, (1) when the high peaks are near one another and hence carry mutual information about their joint locations in genotype space and (2) when parts of the evolving system are quasi-independent of one another... and

hence can be interchanged with modest chances that the recombined system has advantages of both parents.

In other words, evolutionary algorithms are more effective on correlated type-2 functions.

A good summary of the mathematical formulation and characteristics of evolutionary algorithms for real-valued parameter optimization can be found in Fogel [1992]. Regrettably, this article applies the algorithms to type-1 functions, where, as just noted, they have no useful role vis-à-vis traditional algorithms. However, the reported results remain useful for purposes of illustration.

Michalewicz [1996, Chapter 2], gives a very useful and detailed example of an evolutionary algorithm applied to a continuously differentiable type-2 function. An important point to observe in this example, which is typical, is that the *variation* between members of the population of approximate solutions or genotypes *diminishes* as the genotypes maintained by the algorithm converge on a global optimum of F.

Evolutionary approaches can also include *self-adapting* mechanisms for strategy parameters that govern, for example, the mutation operation; and they can be extended to conduct this search systematically within the same optimization framework. Chapter 7 of Bäck [1996] gives a detailed description of this meta-evolutionary approach to obtaining optimal strategy-parameter settings.

Finally, evolutionary algorithms can compute the fitness value of each member of a population in *parallel*, a feature that has been extensively exploited in their implementation; see, for example, Levine [1996].

14.1.3 Optimizing an Optimizer

There is an obvious way to link traditional UM and evolutionary GUM algorithms of the previous two subsections: Use an evolutionary algorithm to seek *optimum* settings for traditional-algorithm parameters. This is well known in the evolutionary algorithms literature. Consider, for example, a Newton–Cauchy algorithm with

- key parameters $\lambda = (\lambda_1, \ldots, \lambda_m)$;
- a type-1 function $f(\mathbf{x})$, $\mathbf{x} \in R^n$, to be minimized;
- a starting point $\mathbf{x}^{(0)}$;
- a convergence test, for example, $\|\nabla f(\mathbf{x})\| \leq \epsilon$, where ∇f, $\|\cdot\|$, and ϵ denote, respectively, the gradient of f, a vector norm, and a small tolerance;
- a performance measure $F(\lambda)$, for example, (a) the number of calls to the function/gradient routine to achieve convergence or (b) the difference between the initial and final function values for a prescribed number of calls to the function/gradient routine.

In general, F will be a type-2 function to which an evolutionary algorithm of Section 14.1.2 can be applied in order to find an "optimal" setting for the parameters, say, λ^*. As one might expect, the function F and its associated λ^* will vary in a drastic and unpredictable manner as one or more of the foregoing quantities that affect $F(\lambda)$ are varied, for instance, f, $\mathbf{x}^{(0)}$, ϵ. Stated informally, each type-1 function f, associated starting point, and termination criterion gives rise to a new type-2 function F (whose local minima over the parameters λ may or may not be correlated). Is it then reasonable to seek *optimal* or "best" choices λ^* for traditional algorithm parameters based on an *explicitly* chosen F or a (relatively small) representative subset of such functions? And even if such a set of F's could be found and a "composite" function derived from them, there is no guarantee that it would be a *correlated* type-2 function, as would be desirable when one is applying a standard evolutionary optimization technique (see the discussion in Section 14.1.2).

To exhibit this phenomenon in more detail, consider the Rosenbrock's function (2.1) and the 4 basic CG algorithms of Section 5.5.1, namely, HS, PPR, DY, and FR, each with the line search accuracy parameter ν, which corresponds to the parameter ACC in the routine LNESRCH of Section 5.4, set to three values 0.1, 0.5, and 0.9. Suppose the resulting 12 algorithms are run from a grid of 25 starting points defined by a square with vertices at the four points $(-10,-10)$, $(-10,10)$, $(10,-10)$, $(10,10)$, with adjacent grid points along each axis separated by 5 units. Within each run, assume that no further periodic restarting of an algorithm is used, for example, after a cycle of $n+1$ iterations. Assume a convergence tolerance of 10^{-3} on the norm of the gradient and an upper limit of 2000 on the number of calls to the function/gradient routine, called f-values henceforth.

The results are shown in Table 14.1. Each line in the table corresponds to a starting point on the grid, which are taken in the following order:

$$(-10,-10),(-10,-5),\ldots,(-10,10),(-5,-10),\ldots,(-5,10),\ldots,(10,10).$$

The algorithm with the smallest number of f-values is shown in boldface on each line of the table, and the symbol * indicates that the upper limit of 2000 is exceeded. Similar results are tabulated in Table 14.2, again using the 25 starting points, but now with 12 randomly chosen triples (λ,μ,ν) from the CG family in (5.10), where values of λ and μ are restricted to be in the (discrete) set $\{0,0.1,0.2,\ldots,1.0\}$ and ν in the set $\{0.1,0.2,\ldots,0.9\}$. The (λ,μ,ν) values are given at the top of Table 14.2.

Each line in the tables corresponds to a type-2 function[1] in the (λ,μ,ν) space. In agreement with conventional wisdom, the tables show that the

---

[1]**Comment on Notation**: In general, the parameters of a type-2 function F are denoted by the vector $\lambda = (\lambda_1,\ldots,\lambda_m)$, as in Section 14.1.2. For the CG illustration, however, this particular parameter vector is denoted by (λ,μ,ν), where the first parameter is, of course, obviously a scalar quantity.

TABLE 14.1. The 12 basic CG algorithms for 25 starting points.

HS 0.1	HS 0.5	HS 0.9	PPR 0.1	PPR 0.5	PPR 0.9	DY 0.1	DY 0.5	DY 0.9	FR 0.1	FR 0.5	FR 0.9
106	203	175	117	**59**	187	*	1660	*	*	1611	*
177	190	108	140	*69*	*66*	158	*79*	197	1064	**57**	**57**
152	157	135	*72*	*77*	179	**52**	**52**	184	132	101	1404
149	*138*	*142*	*183*	*170*	**132**	1786	1723	*	*	*	*
142	**79**	166	*102*	147	155	1028	418	1736	869	1187	1097
452	**68**	**68**	*79*	221	250	149	*	*	160	*	*
37	*43*	*43*	106	158	141	90	129	*	119	73	112
26	*29*	*33*	*25*	**22**	36	69	151	94	73	68	*
33	56	51	58	59	62	137	223	208	108	283	*
136	*155*	*171*	*143*	*189*	*182*	*	*	*	*	*	1607
68	64	**40**	*58*	74	*54*	132	298	288	*46*	278	89
68	*64*	*60*	*58*	*74*	**54**	121	84	244	88	135	243
69	72	*51*	73	*53*	72	114	*58*	206	*57*	**47**	**47**
71	*68*	**51**	*64*	79	*69*	*55*	*58*	497	114	*62*	100
71	*62*	*66*	**59**	*65*	*81*	*80*	190	89	*70*	127	*77*
368	127	119	**58**	174	144	*75*	*	*	97	*	*
98	114	*100*	178	192	195	**69**	*88*	955	152	336	270
124	*105*	*109*	*99*	**86**	*106*	375	365	582	335	401	*118*
102	99	89	72	**47**	95	834	1262	880	807	1223	1929
126	208	190	**100**	*138*	203	914	1117	1528	1396	*	*
128	147	145	**51**	116	176	*	423	592	*	196	*
137	206	154	*65*	77	96	174	113	116	*	**50**	**50**
213	149	155	**75**	*91*	127	306	451	1141	313	354	*
133	165	154	140	**92**	*119*	511	452	542	644	453	1267
155	*152*	*173*	211	**138**	*167*	*	*	*	*	*	*

PPR algorithm with an accurate line search is a good overall choice, even though it is not provably convergent in the absence of restarts (see Powell [1984]). But no single algorithm performs well across the board. Referring again to Table 14.1, let S_j be the set of algorithms corresponding to the grid starting point j for which the number of f-values lies within, say, 50% of the best value on that line. (This set of numbers, on line j, are shown in italics and are within 50% of the smallest number, which is shown in boldface.) It is evident that the intersection of the sets S_j is empty. However, it is easy to see that a small subset consisting of 3 of the 12 algorithms of Table 14.1 performs well over all the starting points (at least one algorithm in the subset has a performance-count entry in boldface or italics on every line of the table). Similar claims hold for the population of 12 algorithms of Table 14.2, and also for Tables 14.1 and 14.2 considered jointly over the population of 24 algorithms. *There is no ideal CG algorithm* in the tables. A relatively small subset of CG algorithms, on the other hand, performs

TABLE 14.2. The 12 randomly selected CG algorithms for 25 starting points.

λ	0.4	0.1	0.7	0.5	0.4	0.9	0.9	0.3	0.3	0.1	1.0	0.1
μ	0.0	0.3	0.3	0.5	1.0	0.8	0.9	0.0	0.7	0.9	0.7	0.3
ν	0.1	0.3	0.9	0.1	0.6	0.1	0.4	0.2	0.1	0.5	0.1	0.7
	54	211	194	183	209	111	333	115	135	144	*	157
	90	235	*134*	*92*	195	239	143	*111*	*117*	164	*108*	223
	104	95	161	121	92	**41**	97	*59*	160	264	86	73
	153	*164*	196	**129**	*190*	353	290	*140*	*176*	231	*	*164*
	154	*149*	211	*136*	**132**	220	209	*139*	*140*	223	928	221
	318	211	119	196	126	231	**76**	113	126	184	244	238
	105	153	170	*93*	*113*	*123*	236	**91**	*101*	288	144	154
	61	*45*	100	66	*60*	98	*59*	*47*	*52*	**43**	73	*45*
	109	**46**	93	92	82	108	114	66	72	115	129	84
	78	148	183	136	224	426	180	134	161	186	*	135
	62	**39**	102	83	64	192	*57*	75	73	*53*	67	**39**
	69	**44**	80	56	91	129	87	69	73	*47*	191	**44**
	65	*49*	75	58	**36**	68	81	66	62	70	*49*	59
	69	**46**	79	89	*60*	174	79	75	72	*47*	*58*	*52*
	64	*51*	95	92	*59*	173	93	*70*	*72*	**48**	88	*65*
	253	201	227	*113*	**83**	178	200	135	*85*	243	183	191
	166	159	*130*	*137*	159	*104*	153	143	*137*	298	**101**	185
	104	*103*	159	*96*	*83*	178	149	*83*	**73**	108	344	*95*
	150	*103*	*107*	188	124	258	222	*108*	120	*103*	814	**77**
	169	*181*	*191*	**144**	*173*	234	353	*151*	*185*	*150*	926	*157*
	79	201	242	147	201	198	231	143	**59**	103	*	205
	88	173	*95*	*90*	135	132	257	**86**	98	166	132	182
	71	108	192	*94*	128	194	288	127	140	224	312	*86*
	141	*137*	**94**	186	*139*	194	182	159	159	229	711	195
	223	*166*	*176*	*200*	*202*	302	268	*152*	**151**	200	*	*211*

in an efficient and robust manner across the set of starting points used in Tables 14.1 and 14.2.

14.2 Population-Thinking and Multialgorithms

The Nobel Laureate Gerald Edelman makes the following observation in his landmark monograph (Edelman [1992, p. 73]):

> It is not commonly understood that there are characteristically biological modes of thought that are not present or even required in other sciences. One of the most fundamental of these is *population-thinking* developed largely by Darwin. Population-thinking considers *variation* not to be an error but, as the great

> evolutionist Ernst Mayr put it, to be real. Individual variance within a population is the source of diversity on which natural selection acts to produce different kinds of organisms. This constrasts starkly with Platonic *essentialism*, which requires a typology created from the top down.

The multialgorithms approach introduced here is based on this fundamental Darwinian idea of population-thinking. It rejects the notion of "best" algorithm, the "essentialist" view. Instead, the *variation* that enters into the key quantities λ, as in Section 14.1.3, that define a population of traditional algorithms (for finding a local minimum of a function f) is treated as *fundamental.* It is this population *taken as a whole*—the variation between its members is desirable and should be preserved—that addresses the minimization of $f(\mathbf{x})$. In contrast to the well-defined quantity $f(\mathbf{x})$, the "value," say "$F(\lambda)$," of an associated "function," say "F," in a multialgorithm is only implicit, within the competition, *in parallel*, between the members of a given population of algorithms. In other words, seeking to explicitly define and optimize "$F(\lambda)$" as in Section 14.1.3 is not viewed as being meaningful, nor is the issue of whether its "local minima" are correlated of any special relevance.

The multialgorithms approach brings alternative evolutionary-biology-based viewpoints to bear on optimization from those used in current evolutionary algorithms. It takes its metaphor from *population-thinking and variation in their broadest sense* (Mayr [1982], [2001], Edelman [1992], Gould [1996]). In contrast, evolutionary approaches of Section 14.1.2 take their metaphor much more directly from the *genetic operations* of mutation/recombination of members of a population (of traditional algorithm parameter values λ in the present context) whose fitness, relative to *every* other potential member of the population, is often assumed to be quantifiable in an objective and explicit manner via a rugged fitness landscape. (Other physical processes, for example, annealing and spin-glasses, have also provided useful algorithmic metaphors for optimization, and in like manner, the basic *physical operations* themselves are mimicked or simulated in such algorithms; see Bounds [1987] for a survey.) It is this excessive reliance on the metaphor at the level of the operations themselves that has been decried by Fox [1995]; see the quotation from this paper near the beginning of Section 14.1.2.

We now give a detailed case study of the multialgorithms idea in the next section, which is followed by a more general discussion in Section 14.4.

14.3 CG Multialgorithms: A Case Study

The key ideas underlying the multialgorithms paradigm are introduced within the specific context provided by the conjugate gradient (CG) method for nonlinear unconstrained minimization. Algorithms derived from it are among the most simple and elegant available in optimization, and they are often surprisingly efficient in practice. Recall, in particular, the four basic CG algorithms introduced in Section 5.5.1 and the two-parameter family (5.10). Henceforth, the subscript k will be dropped in this expression, because the choices for the parameters will not vary with the iteration number.[2]

A traditional optimization approach would seek the "best" values for λ and μ and an associated CG algorithm. Let us now, instead, adopt a nonessentialist, *population-based* perspective.[3]

14.3.1 Fixed-Population Multialgorithms

Consider a finite, fixed *population* of CG algorithms from (5.10). For example, if the four combinations of extreme values for the parameters λ and μ are chosen, each using the three common choices for ν in the line search as in Section 14.1.3, then we obtain 12 possible algorithms. Alternatively, the 12 triples (λ, μ, ν) could be chosen randomly with λ and μ in $[0, 1]$ and ν in $[0.1, 0.9]$. Each triple (λ, μ, ν) will be called the G-type of the corresponding algorithm. The embodiment of the algorithm itself, whose parameters are defined by a G-type, will be called its corresponding P-type. (The motivation for these terms comes from "genotype" and "phenotype.") Let us also include the steepest-descent algorithm ($\beta_k \equiv 0 \ \forall \ k$ in expression (5.10)) with $\nu = 0.1$, leading to a total population of $n_P = 13$ algorithms.

Each algorithm is run in parallel from a given starting point, say $\mathbf{x}$ (the initial direction is along the negative gradient) and is allowed up to M_F calls to the function/gradient evaluation routine that returns information in the standard way; henceforth we say f-value for each such evaluation. (The multialgorithm is terminated if any member finds the solution.) We call M_F f-values the *life span* of one generation of the foregoing population. The best iterate over the population; i.e., the one with the least function value ("fittest" member) can then be identified. If each member of the population is run in parallel, then a *processor group* of n_P processors would be required. In addition to a single generation, at the same time consider another optimization path with two generations, each with a solution life span of $\frac{1}{2}M_F$ f-values. In this case, all members of the population can again be run in parallel for up to $\frac{1}{2}M_F$ f-values, the best iterate, say $\mathbf{x}_+$, and its associated direction $\mathbf{d}$ that produced it taken as the initiating information, and the procedure then repeated for the second generation. This requires a second processor group of n_P processors. In a similar vein, consider ad-

[2] See also the comment on notation of Section 14.1.3: footnote 1.

[3] Indeed, looking in this direction, the generalization to obtain a new two-parameter family of CG algorithms (5.10) became immediately apparent, and this was, in fact, the way the family was originally discovered.

ditional cases of $3, \ldots, n_G$ generations. A total of $n_G * n_P$ processors in n_G processor groups are required.

For example, let us take $M_F = 60$ and $n_G = 5$. This results in the following five cases: one generation with a life span of 60, two generations each with a life span of 30,..., five generations each with a life span of 12. (Alternatively, the different life spans could differ by a fixed quantity, for example, 10, with the last of a generational sequence being truncated so that the total allowed to any processor group in one major cycle is M_F.) When the population is defined by the 13 algorithms chosen earlier, then 65 processors are needed. At the end of a major cycle, the best iterate *over all processor groups* is identified (denote it again by $\mathbf{x}$), and the entire procedure is repeated.

The foregoing is an example of a multialgorithm, and it is summarized by the pseudocode of Figure 14.1. Each major cycle uses a restart at the

0: Given the optimization problem

0: Select an initial population

0: Specify the starting point $\mathbf{x}$

0: sequential_for until the solution is found

major cycle:

1: parallel_for each processor group (PROG)

2: initiate at $\mathbf{x}$ (set $\mathbf{x} \rightarrow \mathbf{x}_+$ and $\mathbf{0} \rightarrow \mathbf{d}$)

2: sequential_for each successive generation of PROG

minor cycle:

3: parallel_for each member of the population

4: run the CG routine for the life span of PROG starting from current $\mathbf{x}_+$ and $\mathbf{d}$ and if it converges then stop

3: end_parallel_for population

3: set $\mathbf{x}_+$ to the best iterate over population and set $\mathbf{d}$ to the associated direction

2: end_sequential_for minor cycle

1: end_parallel_for PROG

1: find the best iterate over all processor groups and set to $\mathbf{x}$.

1: **option:** evolve a new population (see Sec. 14.3.2)

0: end_sequential_for major cycle

FIGURE 14.1 Pseudocode of a CG multialgorithm.

current iterate $\mathbf{x}$. Each minor cycle (after the first in its sequence) is initiated with the best iterate over the population and the (associated) search direction leading to it.

14.3.2 Evolving-Population Multialgorithms

The three quantities (λ, μ, ν) that define the G-type can each take on continuous values in the [0, 1] range. A key option in the multialgorithm of Figure 14.1 is to begin with some initial population (whose size n_P is governed by the number of available processors and whose members are appropriately specified) and to introduce evolutionary operations of *recombination* and *mutation.*

For example, suppose the initial population is chosen using any mix of selected and random values for the parameters with $n_P = 13$. Take $n_G = 5$ and again organize the multialgorithm as in Figure 14.1. At the end of a major cycle, the winner from each of the five processor groups is identified and its G-type retained. A winner is defined to be the G-type that wins most frequently within the set of minor cycles that constitute the major cycle. Another five "descendants" are generated by taking random pairs ("parents") among the winners and then randomly mixing their G-types. An additional set of three "mutants" is obtained either by choosing their G-types randomly (within the appropriate ranges) or by making relatively small random perturbations of G-types that are themselves randomly selected from the set of parents and descendants.

14.3.3 Computational Experiments

We now give a computational demonstration of the foregoing multialgorithms. For present purposes, a parallel machine—typically an MIMD machine or a cluster of PCs or workstations—is not needed, because it is easy to simulate its operations using a single-processor desktop computer, and then to record performance in parallel-machine terms.

The pseudocode of Figure 14.1 was implemented in this manner in Fortran-77 for the two specific multialgorithms discussed earlier in Sections 14.3.1 and 14.3.2, which are summarized in Table 14.3. In all cases, $n_P = 13$ and $M_F = 60$, with five processor groups ($n_G = 5$) whose life spans are as follows: 60, 30, 20, 15, and 12.

The resulting multialgorithms were run on four standard test problems from their standard starting points. These problems and points are defined in Moré, Garbow, and Hillstrom [1981]. (For present purposes of illustration, problems of high dimension are not needed.) The convergence criterion was to terminate when the gradient norm fell below a small prescribed threshold, $\epsilon = 10^{-3}$. A smaller number was not used, because our primary concern here is measuring global rather than asymptotic efficiency, and evaluating robustness. Performance was measured *in parallel machine*

TABLE 14.3. Fixed and evolving-population CG multialgorithms.

FP	Fixed population with λ and μ taking on their extreme values 0 and 1, and with three choices 0.1, 0.5, and 0.9 specified for ν. The last algorithm is steepest descent with $\nu = 0.1$. See Section 14.3.1.
EP	Evolving population, which is chosen initially as follows: The first four members are HS, FR, PPR, and DY with $\nu = 0.1$; for the remaining nine initial members, the three parameters are chosen randomly with λ and μ taking values in the (finite) set $\{0.0, 0.1, 0.2, \ldots, 0.9, 1.0\}$ and ν in $\{0.1, 0.2, \ldots, 0.9\}$. The evolutionary strategy is described in Section 14.3.2. The first of the two options for the mutation strategy is implemented, the values of the parameters again being restricted to the foregoing finite sets.

terms as follows: suppose the number of major cycles taken was N_C and the number of f-values in the last major cycle (terminated at level 4 in Figure 14.1) was N_L. Then the measure of performance was taken to be

$$(N_C - 1) * M_F + N_L. \tag{14.1}$$

We shall say that expression (14.1) gives the number of parallel f-values—henceforth *♯f-values*—used by the multialgorithm.

For purposes of comparison, the twelve individual CG algorithms that are used within the multialgorithm labeled FP in Table 14.3; namely, the HS, FR, PPR, and DY algorithms with three different line search accuracies were each run from the standard starting points to completion (without restarts). Their performance was measured in the usual *single-processor terms*[4] as the total number of f-values to reach the same prescribed accuracy ϵ as the multialgorithms. The results are summarized in the upper part of Tables 14.4 to 14.7 for four test problems: extended Rosenbrock, Brown–Dennis, Watson, and Biggs EXP6 (see Moré et al. [1981] for details). The number of failures (CG routine exceeded 2000 f-values) is reported as #fail. For the *successful* runs, the minimum and maximum f-values and the first, second (median), and third quartiles are given, along with the corresponding CG algorithm and its G-type (λ, μ, ν). The quartiles are denoted by Q1, M, and Q3, respectively. (If a quartile is determined by two numbers, then they are both given along with the corresponding pair of G-types; this will happen when #fail is an even number as in Table 14.6.) For example, for the extended Rosenbrock function of Table 14.4, the Hestenes–Stiefel algorithm with line search accuracy parameter 0.9 took the lowest number of calls (73 f-values), and a quarter of the set of algorithms needed no more than 90 f-values.

[4]This corresponds to (14.1) with a very large value for M_F and thus $N_C \equiv 1$.

TABLE 14.4. Extended Rosenbrock ($n = 10$, $\epsilon = 10^{-3}$).

	f-values	Traditional CG(λ, μ, ν)
Max	1074	DY(1, 0, 0.5)
Q3	619	DY(1, 0, 0.1)
M	108	HS(0, 0, 0.5)
Q1	90	PPR(0, 1, 0.9)
Min	73	HS(0, 0, 0.9)
#fail	1	

Multialgorithm	♯f-values	Major Cycles
FP	55	1
EP	[58, 63, 76]	[1, 2, 2]
#fail	0	

For the two multialgorithms of Table 14.3, the corresponding results, *as measured by the quantity (14.1)*, are tabulated for the four problems in the lower portion of Tables 14.4 to 14.7. Thus the fixed population multialgorithm FP applied to the Brown–Dennis function in Table 14.5 terminated during the third major cycle ($N_C = 3$) and required 131 ♯f-values. Recall that $M_F = 60$, so some algorithm in the fixed population terminated at the optimum during the third cycle after using $N_L = 11$ f-values, and during each of the first two major cycles, *every* algorithm of the population used $M_F = 60$ f-values.

TABLE 14.5. Brown–Dennis ($n = 4$, $\epsilon = 10^{-3}$).

	f-values	Traditional CG(λ, μ, ν)
Max	743	DY(1, 0, 0.9)
Q3	372	FR(1, 1, 0.9)
M	300	HS(0, 0, 0.5)
Q1	188	DY(1, 0, 0.1)
Min	168	FR(1, 1, 0.1)
#fail	1	

Multialgorithm	♯f-values	Major Cycles
FP	131	3
EP	[123, 143, 161]	[3, 3, 3]
#fail	0	

TABLE 14.6. Watson ($n = 9$, $\epsilon = 10^{-3}$).

	f-values	Traditional CG(λ, μ, ν)
Max	668	PPR(0, 1, 0.1)
Q3	[310, 265]	PPR(0, 1, 0.5); HS(0, 0, 0.5)
M	[263, 223]	PPR(0, 1, 0.9); DY(1, 0, 0.1)
Q1	[222, 222]	HS(0, 0, 0.1); HS(0, 0, 0.9)
Min	211	FR(1, 1, 0.1)
#fail	4	

Multialgorithm	♯f-values	Major Cycles
FP	120	2
EP	[107, 153, 189]	[2, 3, 4]
#fail	0	

To capture the variability in performance of multialgorithm EP for different choices of initial population and evolutionary strategy, the technique for choosing random numbers was reseeded 13 times, and the minimum and maximum performance measures were recorded along with the median value. These are the three numbers associated with the multialgorithm EP entry in the lower part of the four tables. They condense the results of these 13 separate runs on each test problem. The corresponding numbers of major cycles that are required are also reported. For example, on Watson's function (Table 14.6), 7 of the 13 reseeded runs of the multialgorithm EP took at most 3 major cycles and used no more than 153 ♯f-values as

TABLE 14.7. Biggs EXP6 ($n = 6$, $\epsilon = 10^{-3}$).

	f-values	Traditional CG(λ, μ, ν)
Max	1822	DY(1, 0, 0.9)
Q3	764	FR(1, 1, 0.5)
M	177	HS(0, 0, 0.9)
Q1	117	HS(0, 0, 0.5)
Min	44	PPR(0, 1, 0.1)
#fail	1	

Multialgorithm	♯f-values	Major Cycles
FP	44	1
EP	[44, 44, 44]	[1, 1, 1]
#fail	0	

defined by expression (14.1). The worst performance of the multialgorithm required 4 major cycles and 189 ♯f-values. The best sequential performer in the upper part of this table was Fletcher–Reeves with an accurate line search, requiring 211 f-values. For this problem, the worst CG algorithm turned out to be the widely recommended choice, PPR with an accurate line search, requiring 668 f-values.

Finally, in Table 14.8, we report results for a larger test problem, the extended Rosenbrock function used in Table 14.4 with $n = 50$. The starting point used was given by

$$\begin{bmatrix} x_{2i-1} \\ x_{2i} \end{bmatrix} = \begin{bmatrix} -1.2 * r_{2i-1} \\ 1.0 * r_{2i} \end{bmatrix}, \qquad i = 1, \ldots, 25,$$

where $r_j, j = 1, \ldots, 50$ are random integers in the range [1, 10]; i.e., the standard starting point for this problem was used with each component scaled by a random integer in the indicated range.

TABLE 14.8. Extended Rosenbrock's function, $n = 50$.

M_F	PROG Life spans	FP	RP	EP
60	(60, 30, 20, 15, 12)	42, **2493**	30, **1743**	30, **1797**
120	(120, 60, 40, 30, 24)	13, **1543**	15, **1773**	14, **1642**
180	(180, 90, 60, 45, 36)	8, **1332**	7, **1227**	7, **1253**
240	(240, 120, 80, 60, 48)	7, **1474**	7, **1563**	8, **1683**
300	(300, 150, 100, 75, 60)	6, **1503**	6, **1711**	6, **1694**
360	(360, 180, 120, 90, 72)	5, **1447**	5, **1578**	5, **1705**
420	(420, 210, 140, 105, 84)	4, **1368**	5, **1755**	4, **1623**
480	(480, 240, 160, 120, 96)	4, **1506**	3, **1181**	3, **1188**
540	(540, 270, 180, 135, 108)	3, **1430**	3, **1209**	3, **1220**
600	(600, 300, 200, 150, 120)	3, **1450**	3, **1211**	3, **1214**
660	(660, 330, 220, 165, 132)	3, **1450**	2, **1268**	3, **1324**
720	(720, 360, 240, 180, 144)	3, **1534**	2, **1217**	2, **1184**
780	(780, 390, 260, 195, 156)	3, **1563**	2, **1263**	2, **1288**
840	(840, 420, 280, 210, 168)	2, **1562**	2, **1176**	2, **1236**
900	(900, 450, 300, 225, 180)	2, **1513**	2, **1247**	2, **1299**
⋮	⋮	⋮	⋮	⋮
1800	(1800, 900, 600, 450, 360)	1, **1537**	1, **1279**	1, **1279**
$n_G = 1$		1, **1849**	1, **1626**	1, **1626**
G-type		(0, 0, 0.5)	(0.1, 0.3, 0.3)	(0.1, 0.3, 0.3)
	HS	**2212**		
	FR	**≥ 3000**		
	PPR	**1900**		
	DY	**≥ 3000**		

A number of different values of M_F were used, as shown in the first column of Table 14.8. For each value of M_F, the life spans for the $n_G =$ 5 processor groups are M_F, $M_F/2, \ldots,$ M_F/n_G, respectively. These are given in the second column of the table. Each subsequent column gives the number of major cycles and the number of ♯f-values (in boldface) for a multialgorithm. Also, an additional multialgorithm was run corresponding to the choice of the intial population of the multialgorithm EP of Table 14.3, which was then left unchanged; i.e., the subsequent evolutionary strategy used in EP to change the population between major cycles was omitted. Results for this multialgorithm are reported in the column RP.

In addition, the last line of the upper table shows results with $n_G = 1$ and a very large choice of M_F (machine infinity). This corresponds, in effect, to running each member of the multialgorithm in parallel and choosing the best one as indicated by the corresponding G-type.

The bottom part of the table gives the results of running the four basic CG algorithms with an accurate line search; i.e., $\nu = 0.1$.

The numerical results provide a useful illustration of the performance of the CG-multialgorithms as well as *a platform for the more general discussion* in the next section.

14.4 The Multialgorithms Paradigm

When an optimization technique, for example, a Newton–Cauchy method of Chapter 2, is formulated into an algorithm, certain quantities hold the key to good performance. These parameters are usually few in number. A *population* of algorithms is defined, each member by setting the key parameters to particular values. The premise underlying a multialgorithm is that the *variation* within this population is *real*; i.e., it does not simply represent a departure from some *ideal* algorithm. The underlying motivation and metaphor come from evolutionary biology, as discussed in Section 14.2. We now continue that discussion after introducing some terminology.

14.4.1 Terminology

Motivated by the "population-thinking" terminology of evolutionary biology, but also seeking to maintain a distinction between the names used there and those directly appropriated for evolutionary algorithms, henceforth in the discussion we will use the term *G-nome* to mean the set of key parameters and their numerical range of values that define the population of algorithms. For example, the G-nome of the CG family of Section 14.3 is defined by (λ, μ, ν) with λ and μ in the range $[0, 1]$ and ν in the range $[0.1, 0.9]$. A particular choice of values for the G-nome defines a *G-type* (motivated by "genotype"). For example, PPR with an accurate line search has G-type $(0, 1, 0.1)$. The complete realization or embodiment of the algorithm itself is called its *P-type* (motivated by "phenotype"). The G-type of an algorithm does *not* encode the embodying P-type. (In biology, the genotype of an organism is often interpreted as encoding, in essence, the phenotype, although this is by no means the universal view among biologists; see, for example, Edelman [1988] for an alternative perspective.) Also, *the* population means the entire (possibly infinite) set of possibilities determined by the G-nome, and *a* population means a particular subset of G-types.

14.4.2 Discussion

Tables 14.1, 14.2 and the graphs of Figure 5.1 depict several functions of type 2 as characterized in Section 14.1.2 that are associated with the standard two-dimensional Rosenbrock function f, a stopping tolerance ϵ, and a performance measure defined by the number of f-values used. They are defined over a discrete domain of G-types, and whether they have correlated or uncorrelated local minima is not a concern here. Each such function, say F, comes from changing the starting point. More generally, other entities that determine F, for example, the (type-1) function f, the stopping criterion, the performance measure, could be changed. In a multialgorithm, there is no attempt to optimize a particular (type-2) function, F. (Which candidate F or set of candidates would one choose?) Instead the aim is to capture the variation—within a single F and between different F's—through the use of a relatively small population of algorithms. This population, as a whole, is used to minimize the given type-1 objective function f. Progress is measured in terms of a local and highly context-dependent *ordering* that arises from the competition, in parallel, between members of a particular population as they seek to minimize f. But there is no attempt to identify an *explicit* function F that associates a global, context-free fitness *value* with a given G-type. Indeed, the "fitness" of a winner in the CG-multialgorithm of Section 14.3.1 within a major or a minor cycle emerges as much from its own efficacy during the cycle as it does from its earlier cooperation (with other algorithms in the current population) that yielded the starting point used by the cycle.

As indicated by the numerical experiments of Section 14.3; see the second half of Tables 14.4 through 14.7—a population that is fixed in advance would be adequate. Not much is gained when evolutionary operations are included in the multialgorithm in that *particular* experiment. *It is conjectured that this could be true much more broadly.* However, to cover more complex situations, it is important also to have the option of being able to explore the space of G-types *in order to revise and evolve the population*, and here the considerable body of knowledge concerning evolutionary algorithms can be adapted to present needs. These techniques need *not* be justified by seeking a counterpart in nature; recall the cautionary advice of Fox [1995] quoted in Section 14.1.2. But they are always premised on the population principle, wherein variation of certain key quantities within a population of traditional algorithms is viewed as being of fundamental importance.

Because multialgorithms seek to preserve variation, the way they define genetic-type operations can differ significantly from current evolutionary algorithms. However, ideas that underpin genetic algorithms, for example, *intrinsic* or *implicit parallelism* (as contrasted with the explicit use of parallel computers), the *schemata theorem*, and the identification of *building blocks*, can all still come into play in a significant way; see Holland [1975], [1995].

Like population size, the choice of *life spans* within a multialgorithm is an important issue. A *trivial* multialgorithm can be obtained by taking $M_F = \infty$ and $n_G = 1$ and running each algorithm from a fixed population in parallel from a given starting point. The use of shorter life spans within major and minor cycles becomes natural, both for enhancing efficiency and robustness; see our subsequent discussion on convergence in Section 14.6.1—and as a means for discovering more effective populations via evolution-type operations. Several variants can be envisioned on the schemata shown in Figure 14.1. For instance, instead of life spans being fixed in advance, they can be incorporated into the G-type and themselves subjected to selection.

In developing a theory of multialgorithms, the illustrative CG multialgorithms of Section 14.3 may prove to be useful *models* for capturing some of the essential characteristics of multialgorithms in general. For an analogue, see Kauffman [1993, pp. 40–47], which employs a simple and convenient "NK-model" to capture the essential characteristics of more general "rugged fitness landscapes." We conjecture that a "CG-model" could fulfill a similar role vis-à-vis more general "multialgorithms."

In addition to the two fundamental assumptions of Section 2.1.1, a multialgorithm operates under an additional premise as follows:

- *A population is needed to capture the variation within the G-nome.* The chosen size of this population is a key quantity. Consider the CG multialgorithms of Section 14.3. Suppose λ and μ are taken in increments of 0.1 in the range [0, 1] and ν in similar increments in the range [0.1, 0.9]. Then approximately a thousand possible algorithms (G-types) would result. If they are run in parallel within the multialgorithm framework of Section 14.3, with $n_G = 5$, then approximately 5000 processors would be needed. Little *may* be gained from defining a population with such a fine gradation for the G-nome parameters; i.e., a fairly small number of G-types may be sufficient to capture the desired variation. This is certainly indicated by the computational illustration of Section 14.3, but requires much more investigation before any definite conclusions can be drawn.

14.5 Parallel Computing Platform

A key development, with fundamental implications for optimization, has been the major advances in parallel computation made during the past decade. For example, in describing a parallel computing project at the California Institute of Technology, G.C. Fox, Williams, and Messina [1994] make the following observation:

> Parallelism may only be critical today for supercomputer vendors and users. By the year 2000, however, all computers will have to address the hardware, algorithmic, and software issues implied by parallelism. The reward will be amazing performance and the opening up of new fields, the price will be a major rethinking and reimplementing of software, algorithms and applications.

Distributed memory, multiple-instruction multiple-data (MIMD) parallel machines, computing clusters, and distributed computing systems are especially suited to the optimization context; see, for example, Schnabel [1994]. Affordable desktop machines of these types are close at hand, and good ways to utilize them for optimization will obviously be needed. A dominant approach so far, although not exclusively so, has been that of parallelizing existing, appropriately chosen, sequential algorithms. This has resulted in a revisiting of algorithmic approaches previously consider outdated—for instance, Dennis and Torczon [1991] in the area of derivative-free optimization—but no dramatically new approaches seem to have emerged so far, at least from a conceptual standpoint.

MIMD desktop machines with a significant number of processors that are relatively sparsely interconnected are likely to become inexpensive in the near future. In particular, a Beowulf-type cluster of between 16 and 128 processors, say, is nowadays quite inexpensive to build; see the CACR Annual Report [1997, p. 20]. Such systems exploit commodity silicon technology, for example, Intel PentiumPro processors; commodity networking, for example, fast Ethernet connections; free operating systems, for example, Linux; and industry-standard parallel programming environments, for example, MPI (see Gropp et al. [1994]). These are precisely the types of architectures for which an optimization multialgorithm is well suited. Inexpensive Beowulf and other MIMD machines with hundreds and possibly thousands of processors are on the horizon.

It is evident from Figure 14.1 that the multialgorithm approach is inherently and highly parallel in nature. Indeed, we were able very easily to adapt the CG multialgorithm code of Section 14.3 to the *basic six-function MPI environment*, Gropp et al. [1994, Chapter 3], and execute it on a commercial MIMD computer that supported MPI. The latter—the HP Exemplar at the Center for Advanced Computing Research (CACR), California Institute of Technology—is a high-end machine that was very convenient for carrying out this experiment with MPI parallelization of the CG-multialgorithm code. But it is obviously not the inexpensive environment we have in mind for practical use of a multialgorithm. A detailed exploration of how to exploit a Beowulf cluster or similar inexpensive parallel computer to implement a multialgorithm is left to a future work. See also comments on the Nelder–Mead family of algorithms in Section 14.6.3.

14.6 Other Topics

14.6.1 Convergence Analysis

Within a family of optimization algorithms, for example, the nonlinear CG family of Section 14.3, one algorithm may be *provably convergent* but slow; another may be *fast in practice* but have no associated global convergence proof; and yet a third may have fast *asymptotic* convergence and lack the other two desirable properties. The formulation of an acceptable sequential algorithm may be a complex task involving compromises among the foregoing three objectives.

Multialgorithms provide a new avenue for ensuring convergence without sacrificing efficiency. Not all the algorithms in a population are required to be *provably* convergent. For example, in the fixed-population CG multialgorithm (Section 14.3), most of the constituent algorithms are not known to be convergent. The inclusion of a single, perhaps quite inefficient, algorithm, whose convergence is guaranteed, is all that is required. Thus multialgorithms make it possible to implement "pure" versions of traditional algorithms that are well regarded and obtain a convergence guarantee in a different way. (The algorithms must, of course, be safeguarded against overflows and other potential numerical difficulties, so one member does not cause the entire multialgorithm to fail.) Compromises that are often introduced into traditional algorithms to ensure convergence can be avoided. For example, the PPR algorithm of Section 5.5.1 without restarts is known to cycle on certain problems (Powell [1984]), and it must be modified in order to guarantee convergence. Yet in practice, PPR is often a very efficient member of the nonlinear CG family.

The foregoing claims seem almost self-evident, in particular for the CG multialgorithms of Section 14.3 using an appropriate choice of M_F. But they must be mathematically formalized. For this purpose, the use of point-to-set algorithmic maps and convergence analysis techniques of Zangwill [1969] and Luenberger [1984] can be utilized in order to show formally that a multialgorithm, in a parallel computing environment, can exhibit the best of three worlds: ensured convergence from an arbitrary starting point, good asymptotic rate of convergence, and good overall efficiency or global complexity.

14.6.2 A Performance Evaluation Environment

In the previous subsection we focused on the case where the $f/g/H$ routine is not amenable to parallelization and on the use of available parallel processors to keep the multialgorithm iteration count and the associated number of calls to the $f/g/H$ routine as low as possible. But nothing precludes the converse, where a parallel machine is used, in its entirety, to speed up $f/g/H$ evaluation (and other algorithm overhead) within a single traditional algorithm.

In this context, a multialgorithm approach can provide a built-in *performance evaluation environment*, and thus facilitate the choice of this single algorithm from its constituent family. In a fixed-population multialgorithm, one can determine good choices for the G-type by keeping track of winners; i.e., a good algorithm will emerge during execution. In an evolving-population setting *with an appropriately chosen recombination/mutation strategy*, which does not necessarily emphasize the preservation of variation and is closer to that used in current evolutionary optimization algorithms, the G-types of winners will be replicated, and the emergent population itself will identify the most effective algorithms; i.e., the population of algorithms can serve as a "counter" of success. Once a suitable algorithm is identified, then it can be parallelized in the traditional manner.

14.6.3 Further Areas of Application

Each of the main families of Newton–Cauchy algorithms for unconstrained minimization discussed in earlier chapters is embraced by the multialgorithms paradigm. The key parameters that can be used to define their G-types have been identified there. Most traditional algorithmic areas of nonlinear equation-solving and linear and nonlinear constrained optimization are also embraced by the multialgorithms paradigm. Consider again a few specific examples:

- **L-BFGS**: In recent years, the nonlinear conjugate gradient algorithm has taken a back seat to the *limited-memory BFGS algorithm* for practical applications; see Section 2.3.3. Key quantities that determine the algorithm's effectiveness are the number of retained step and gradient-change pairs, say t, the choice of initial scaling matrix, and the line search accuracy. (The choice $t = 5$ for the first parameter and the identity matrix with Oren–Luenberger sizing for the second parameter are recommended in practice; see Gilbert and Lemaréchal [1989], but these are only rules of thumb.) L-BFGS is an excellent candidate for formulation as a multiagorithm, with its G-nome defined in terms of the foregoing quantities.
- **LG**: In the *Lagrangian globalization* method of Section 3.3 for solving a system of nonlinear equations, the commonly used sum-of-squares

merit function is replaced by a Lagrangian-based potential. The latter contains parameters that are difficult to choose a priori without substantial knowledge about a given problem, in particular, bounds on higher-order derivatives in a region of interest. The G-type of a multialgorithm could include these parameters, along with others governing the line search accuracy and the perturbation strategy in ill-conditioned regions; see Nazareth [1996a]. Also, mixed strategies developed in Nazareth and Qi [1996], which interleave the sum-of-squares merit function and the Lagrangian-based potential function in order to enhance robustness without sacrificing efficiency, can be replaced by *pure* strategies (basic forms of the algorithms) that are run, in parallel, within the framework of a multialgorithm.

- **SQG**: The *stochastic quasi-gradient method* is a version of Cauchy's method—see also Chapter 9 on the quasi-Cauchy method—for optimization under uncertainty that has been shown to be very useful for a broad range of practical applications; see Ermoliev and Gaivoronskii [1994] and other articles of these authors in Ermoliev and Wets [1988]. The step-length strategy, the choice of diagonal scaling matrix, and the manner in which the algorithm is terminated all hold the key to effective performance, and again, only rules of thumb apply when the sequential algorithm is implemented in practice. *The SQG method lends itself in a very natural way to formulation as an SQG multialgorithm*, and the latter is likely to be much more effective in practice.
- **NM**: The Nelder–Mead simplex method—see Chapter 6—is a very natural candidate for a multialgorithm implementation. The main expansion, contraction, and shrink parameters can define the G-nome, and the NM-GS algorithm of Section 6.3 would correspond to a particular choice of G-type. The introduction of fortified-descent, Tseng [2000], can guarantee convergence of any NM-type algorithm on a smooth function. (As discussed in Section 14.6.1, only a single G-type within the multialgorithm need have this property in order to ensure convergence of the multialgorithm.) In particular, *an implementation of an NM multialgorithm on a Beowulf machine could lead to a very useful tool for minimizing nonlinear functions in the presence of noise.*
- **IPM**: Many *interior-point methods* for linear programming use a search direction that is composed of a combination of up to three key vectors: affine scaling, centering, and feasibility. For details; see, for example, Vanderbei [1996] or Nazareth [1996b], [1998]. The G-type of a multialgorithm could include parameters that define the linear combination, the step length along a search direction, and, if inexact computation is used, the accuracy to which the component vectors are computed. A potential function could determine fitness. Many studies of individual algorithms based on sophisticated strategies for com-

bining the three vectors can be found in the literature. It would be interesting to explore whether much simpler, ad hoc, population-based heuristics, in the setting of parallel computation, would work just as well in practice.

The primary focus in this chapter has been on multialgorithms for optimization. In future work, we will focus, at first, on the development of the NM and SQG multialgorithms outlined above and their implementation on a Beowulf machine.

The ideas introduced here also apply to other areas of scientific computing, for instance, homotopy methods for nonlinear equation-solving; see Keller [1978], [1987], Allgower and Georg [1990], Garcia and Zangwill [1981]. Another good starting point can be found in the article of Weerawarana et al. [1996].

14.7 Recapitulation

Carrying the metaphor from evolutionary biology one further step, we may observe that the notion of "species" arises quite naturally within the formulation of a multialgorithm. For example, in the illustration of Section 14.3, the steepest-descent algorithm is a different "species" of algorithm that is used to augment the fixed population of CG algorithms. Other types of CG-related algorithms can be included, for example, limited-memory BFGS algorithms that use minimal storage; see Chapter 2. These algorithm species, each with its associated G-nome definition, could be incorporated into a single evolutionary multialgorithm and selected for fitness against one another. Recombination operations would be valid only between members of the same algorithm species, and thus different species would remain distinct. In a given problem environment, one algorithm species may gradually displace another, or two or more algorithm species may find it beneficial to cooperate. In this interpretation, the previously mentioned BFGS-CG relationship, which originated in mathematical/physical notions of continuity and conservation; see, in particular, the discussion in Chapter 13—can now be viewed from an alternative biologically based standpoint, as the connection between two interrelated algorithm species. This example highlights the *paradigm shift* advocated in the present discussion. For another example, consider the huge proliferation of algorithms based on an NC "body plan" as summarized in Section 2.4.2 and Figure 2.2, which is again strangely reminiscent of the situation in biology.

Traditional gradient-related methods for optimization over the real field are often named for *Newton*, who invented the differential calculus, and for *Cauchy*, who discovered the method of steepest descent. Both techniques underpin methods for smooth optimization. In an analogous way, it seems reasonable to associate the name of *Darwin* with techniques for minimizing real-valued functions that are *population-based.* In particular, the techniques developed in this article could appropriately be called *Darwinian multialgorithms.* An optimization software library for a parallel machine, whose individual routines implement different Darwinian multialgorithms, could have an added layer that makes its possible to use a set of such routines *simultaneously* to solve a particular problem. Taking the above notion of algorithm species one step further, various "multialgorithm taxa" would compete against or cooperate with one another, and the most effective multialgorithm or combination of multialgorithms would emerge. The software library could thus *adapt* to the problem. For example, it might discover, for a given problem, that an effective strategy is to use a simple multialgorithm to traverse a noisy region and then switch to a more expensive multialgorithm with fast asymptotic convergence, the transition from one to the other emerging during the course of usage.

Darwinian multialgorithms have characteristics that are reminiscent of their counterparts in evolutionary biology. Consider, for example, the following: the need to maintain algorithm-population variation or diversity; the marked distinctions between G-type and P-type; the definitive role played by a small number of G-nome parameters whose values distinguish similar P-types; the way particular parameter settings ("genes") can proliferate rapidly in an evolving population of algorithms; the displacement of one algorithm species by another; the possibility of symbiotic cooperation betweeen them; the proliferation of different algorithm species; the grouping of algorithm species into "taxa." This suggests that the traffic between biological science and algorithmic science may not necessarily be one way. Darwinian multialgorithms, in particular, the CG multialgorithms formulated in Section 14.3, could provide a useful arena for studying certain issues in evolutionary biology; i.e., a population within a Darwinian multialgorithm could serve, in return, as a metaphor for a population of evolving biological individuals. This is only a conjecture at this stage, but it deserves further investigation.

14.8 Notes

Sections 14.1–14.7: Most of the material in this chapter is derived from Nazareth [2001c].

15

An Emerging Discipline

The study of real-number algorithms for the problems of continuous mathematics has a long history. Many of the great names in mathematics, for example, Cauchy, Euler, Gauss, Lagrange, Newton, are associated with basic real-number algorithms for solving equations or minimizing functions, as we have seen in earlier chapters of this monograph. However, it was only with the advent of the electronic computer and the maturing of the field of computer science that the study of real-number algorithms was able to grow to full stature. The computer helped to foster the development of

- a *core* theoretical discipline, by giving stimulus to the development of appropriate real-number models of computation and, more importantly, the discovery and analysis of fundamental new algorithms;
- the *empirical* side of this subject, by providing a laboratory for experimentation based on powerful machines and high-level languages, for example, Matlab, Mathematica, and Maple;
- the *practical* side, by permitting problems that were previously unapproachable to be solved in a routine manner.

A new discipline has taken shape that has a fundamental *algorithmic* core grounded in continuous mathematics and an empirical and practical aspect characteristic of a *science*. In this concluding chapter we discuss some of the issues that underlie this emerging discipline. We also revisit two key algorithmic distinctions introduced earlier, namely, *conceptual* vis-à-vis *implementable* algorithms (Chapters 3, 5, and 10) and *mathematical* vis-à-vis *numerical* algorithms (Chapter 7).

15.1 Background

The study of algorithms clearly falls within the province of computer science. For example, in an invaluable collection of articles discussing the philosophy of computer science, Knuth [1996, p. 5] makes the following observation:[1]

> My favorite way to describe computer science is to say that it is the study of algorithms.... Perhaps the most significant discovery generated by the advent of computers will turn out to be that algorithms, as *objects of study*, are extraordinarily rich in interesting properties; and, furthermore, that an algorithmic point of view is a useful way to organize information in general. G.E. Forsythe has observed that "the question "What can be automated?" is one of the most inspiring philosophical and practical questions of contemporary civilization."

However, one need only look at one of the standard computer science texts, for example, the multivolume classic starting with Knuth [1968], or a thematic overview of the subject, for example, Hopcroft and Kennedy [1989], to see that the emphasis in computer science is primarily on algorithms that arise within the construction and maintenance of the tools of computing; i.e., *algorithms of discrete mathematics*, and not on algorithms that arise when computers are used for real-number computation. Indeed, in the above-quoted work, Knuth notes the following:[2]

> The most surprising thing to me, in my own experiences with applications of mathematics to computer science, has been the fact that so much of the mathematics has been of a *particular discrete type*....

One can find these typical mathematical needs of computer science gathered together in Graham, Knuth, and Patashnik [1989] under the rubric *concrete mathematics*, where "*con*" stands for continuous and "*crete*" for discrete, and where the latter predominates.

We see that despite the early, de jure goals of computer science to be more all-embracing, a *de facto alliance has developed between computer science and discrete mathematics*, underpinned by the Turing-machine complexity model (Garey and Johnson [1979]). Increasingly close ties have been forged, in recent years, between academic departments of computer science and computer engineering under the banners EECS or CS&E, and these too have contributed to the separation of computer science from continuous mathematics. Thus, in an apt and prescient definition of Newell, Perlis, and

[1] Italics ours.

[2] Italics ours.

Simon [1967], modern computer science is best characterized, nowadays, simply as *"the study of computers"* and the phenomena surrounding them in all their variety and complexity. The study of discrete algorithms is one of these phenomena, which lies at the core of the discipline of computer science and is sufficiently rich to unify its various branches.

A fundamental gap has opened between the study of algorithms of discrete mathematics within computer science and the study of real-number algorithms of continuous mathematics. Problems that require real-number algorithms and associated software for their solution arise in all areas of science and engineering, and especially within the fields of *numerical analysis* and *optimization.* Algorithmic numerical analysis, which centers on the solution of systems of algebraic equations and/or ordinary and partial differential equations, used to find a home in computer science departments, but researchers in this field have drifted back in recent years from computer science to mathematics departments. Researchers in algorithmic optimization, whose area centers on the task of locating extrema of unconstrained or constrained functions over finite- or infinite-dimensional domains, once found a primary home within departments of operations research. But the latter field is increasingly tied to engineering (industrial and operational) and to management science, and again there has been a drift of optimization researchers back to mathematics departments. Mathematics has been welcoming back these prodigal sons, and mathematics departments are once again beginning to view themselves as pure, applied, and computational in a merging of the constituent disciplines; see, for example, Renegar, Shub and Smale [1996], Trefethen [1992], and Nazareth [1993a].

Further evidence for the separation between algorithmic science and mathematics (AS&M), on the one hand, and computer science and engineering (CS&E), on the other, can be found in the study of the underlying foundations, and, in particular, the development of alternative computational models (to the Turing model) for real-number computation; see the recent landmark work of Blum, Cucker, Shub, and Smale [1998], and also earlier work surveyed in Traub and Werschulz [1998]. In discussing their motivation for seeking a suitable theoretical foundation for modern scientific and engineering computation, which deal mostly with real-number algorithms, the authors of these works quote the following illuminating remarks of John von Neumann, made in 1948:[3]

> There exists today a very elaborate system of formal logic, and specifically, of logic applied to mathematics. This is a discipline with many good sides but also serious weaknesses.... Everybody who has worked in formal logic will confirm that it is one of the technically most refractory parts of mathematics. The reason for this is that it deals with rigid, all-or-none concepts, and

[3] Italics ours.

> has very little contact with the continuous concept of the real or the complex number, that is, with mathematical analysis. Yet analysis is the technically most successful and best-elaborated part of mathematics. Thus formal logic, by the nature of its approach, is cut off from the best cultivated portions of mathematics, and forced onto the most difficult mathematical terrain into combinatorics.
>
> The theory of automata, of the digital, all-or-none type as discussed up to now, is certainly *a chapter in formal logic.* It would, therefore, seem that it will have to share this unattractive property of formal logic. It will have to be, from the mathematical point of view, combinatorial rather than analytical.

We suggest the name *algorithmic science* for the opening of a new chapter rooted in analysis as a counterpoint to the aforementioned "chapter in formal logic." Why a different name when "computational science," "computational mathematics," and "computational engineering" are already in widespread use today; see, for example, Raveché et al. [1987]? Because, strictly speaking, "computational" within these names does not designate a new *discipline* of science, mathematics, or engineering, but rather a third modus operandi, made possible by the computer, that complements the long-established "theoretical" and "experimental" modes. In other words, *every* field within science, mathematics, and engineering today has its theoretical (or pure), its experimental (or applied), and its computational (or constructive[4]) modes of operation. Other names in current use are "scientific computing," which is a catch-all term that is much too broad an umbrella and does not capture a discipline whose primary objects of study are algorithms; and "algorithmic mathematics," which is too narrow, because it leaves out the important aspect of an empirical and practical science. Thus the term algorithmic science, where the focus is on real-number algorithms of continuous mathematics (without necessarily excluding the discrete), seems to best capture the emerging discipline, and it provides an appropriate counterpoint to computer science, where the focus, as we have seen, lies with the algorithms of discrete mathematics (without necessarily excluding the continuous).

15.2 What Is an Algorithm?

Again in the words of Knuth [1996, p. 5]:

> An algorithm is a precisely-defined sequence of rules telling how to produce specified output information from given input infor-

[4]See Goldstein [1967] for a pioneering work in this area.

mation in a finite number of steps. A particular representation of an algorithm is called a program.

Algorithms for solving certain isolated problems date back to antiquity, and the word "algorithm" itself is derived from the name of the ninth century Persian mathematician Al-Khowârizm. Today, the word has entered popular culture, as evidenced by the recent book *The Advent of the Algorithm* by David Berlinski [2000, p. xix], where one finds the following noteworthy characterization:

> **In the logician's voice:**
> an algorithm is
> a finite procedure,
> written in a fixed[5] symbolic vocabulary,
> governed by precise instructions,
> moving in discrete steps, 1,2,3, . . . ,
> whose execution requires no insight, cleverness,
> intuition, intelligence, or perspicuity,
> and that sooner or later comes to an end.

For example, consider the following oft-quoted classical algorithm (Stewart [1987]), dating back to 300 B.C. and named for Euclid, that finds the highest common factor (HCF) of two positive integers A and B:

> Divide the larger of the numbers, say A, by the smaller, say B, to obtain a remainder C. Then replace A by B and B by C and repeat, terminating when the remainder C becomes zero. The HCF is the final value of B.

For two particular inputs, say, 3654 and 1365, this is described by the *recipe* of Figure 15.1 (Penrose [1989), which returns the answer 21 as the HCF.

$3654 \div 1365$	gives remainder	924
$1365 \div 924$	gives remainder	441
$924 \div 441$	gives remainder	42
$441 \div 42$	gives remainder	21
$42 \div 21$	gives remainder	0

FIGURE 15.1 Recipe for the HCF of two positive integers, 3654 and 1365.

[5]Inadvertantly or not, the author does not explicitly state whether the vocabulary is *finite*, *countably infinite*, or *uncountably infinite*. Also, Berlinski's characterization does not adequately distinguish between an *algorithmic process* (dynamical system that need not be teleological, e.g., Darwinian evolution) and an algorithm (constructive $O \leftarrow I$ map as highlighted in the above quotation from Knuth [1996]).

The formulation of an *algorithm* requires a precise specification of the division/remainder operation and the way in which integers are represented. For the latter, any number system, for example, decimal or binary, would suffice. But the simplest approach is to use the unary system, where an integer is represented by a sequence of dots, each of the form •. For example, the integer 4 is represented as a sequence of four •'s. The division/remainder subroutine operation with divisor B then removes B dots at a time until no further set of B dots can be removed. The procedure for finding the HCF described above can then be expressed formally as a flowchart—see Penrose [1989, pp. 32–33]—that can be executed *in a completely mechanical way* to obtain the highest common factor of any two integers A and B. This clearly fits all the requirements of the foregoing characterizations of an algorithm. In particular, the fixed symbolic vocabulary is just the symbol •. The procedure is finite and guaranteed to terminate. It works on *any valid set of inputs* consisting of two positive integers. And an estimate can be made for the amount of effort involved; i.e., the *complexity* of the algorithm, which is easily seen to be polynomial in the length of the input (the number of dots needed to represent A and B).

As illustrated by this simple example, an algorithm solves a *problem class*, not just one particular problem instance as in Figure 15.1. Furthermore, *complexity estimates*, obtained by theoretical and/or experimental means, are essential to an understanding of computational effectiveness. For these reasons, the oft-used analogy of a recipe in a cookbook[6] does *not* adequately capture the idea of an algorithm. Interestingly enough, it was only in the twentieth century that the *concept of a general algorithm* was formulated by Church, Gödel, Kleene, Markov, Post, Turing, and others. The development, by this group of distinguished mathematicians, of several different models of computation, which were later shown to be equivalent to one another, and the study of the fundamental limits of computability are often counted among the great intellectual achievements of the twentieth century.

Among the different approaches, Turing's proved to be preeminent as a *foundation for computer science*. A Turing machine (TM) model is an idealized computer with the simplest structure possible. A TM has an infinite tape divided into square cells, and a head that can read or write on a cell and move left or right. Its basic vocabulary consists of just two symbols, 0 and 1, and its basic set of commands are listed in Table 15.1 (Stewart [1987]). A *Turing–Post* program is a set of numbered steps chosen from the commands in Table 15.1. *An algorithm, in Turing's approach to computability, can be viewed as just such a program that terminates on all inputs.* For example, Euclid's algorithm discussed above can be expressed, albeit laboriously, as a Turing–Post program. For a detailed discussion of

[6]A cookbook recipe has much more the "flavor" of the recipe of Figure 15.1.

TABLE 15.1. Basic commands of a Turing machine.

Write 1 in the current cell (under the head).
Write 0 in the current cell
Move head to the right
Move head to the left
Goto program step k if the current cell is a 1
Goto program step k if the current cell is a 0
Stop

the TM model, see Garey and Johnson [1979], and for descriptions intended for a general audience, see, for example, Stewart [1987, Chapter 19] and Penrose [1989, Chapter 2].

In the digital Turing machine setting, the computer science view of numerical computation is expressed in Knuth [1996] as follows:

> Computing machines (and algorithms) do not only compute with *numbers*. They deal with information of any kind, once it is represented in a precise way. We used to say that a sequence of symbols, such as a name, is represented inside a computer as if it were a number; but it is really more correct to say that a number is represented inside a computer as a sequence of symbols.

In other words, paraphrasing the well known declaration regarding "taxation," the unspoken mantra of computer science is that there can be

No (numeric) *computation without* (finite, symbolic) *representation*!

15.3 Models of Computation

Let us now consider models of computation that are more naturally suited to building a foundation for algorithmic science.

15.3.1 Rational-Number Model/RaM

The *rational-number model* and the associated idealized machine, known as a rational-arithmetic machine[7] or RaM, are standard vehicles for the study of computability and complexity of algorithms for solving problems of science and engineering. Each "word" of a RaM holds a *pair* of integers,

[7] We use the acronym RaM, because RAM is currently used for "Random Access Machine" in computability theory.

typically in the binary number system that thereby defines the fixed, finite vocabulary for the RaM, and the (implicit) ratio of the two integers defines the represented rational number. Note that the integers can be of arbitrary size, and thus, like the tape of a Turing machine, each word is arbitrarily large. The basic rational-arithmetic operations, $+$, $-$, $\times$, and $\div$, are implemented in the standard way as operations involving integers, whose execution can be specified in a mechanical manner requiring no expertise, as befits the characterizations of an algorithm of the previous section.

It is well known that a RaM can be emulated[8] on a (digital) Turing machine and that the number of TM steps to perform *each basic arithmetic operation* of the RaM is bounded by a polynomial function in the number of bits needed to store the two operands involved in that operation.

15.3.2 Real-Number Model/GaM

A *real-number model* has been formulated in the recent landmark work of Blum, Cucker, Shub, and Smale [1998]. (The associated machine is identified by the acronym BCSS.) The survey monograph of Traub and Werschultz [1998, Chapter 8] provides a comparison of the Turing/RaM and real-number models. For antecedents more directly in the Turing tradition, see, for example, Abramson [1971] and other references cited in the above works.

In real-number models and associated machines, an *uncountably infinite vocabulary* is used, namely, the real numbers. To give the discussion a more concrete form, let us introduce a real-number counterpart of a RaM, which we will call a Greek-arithmetic machine, henceforth abbreviated by GaM. Its machine "word" stores a *geometric symbol*, namely, a rod of infinitesimal width[9] whose *length* equals the real number represented, and whose "color," say black or red, identifies whether the number is positive or negative. The number zero is represented by the absence of a symbol (a blank). The operations $+$ and $-$ between two positive or two negative numbers (symbols of the same color) are performed *geometrically* by aligning the two component symbolic rods lengthwise and producing from them a new geometric symbol of the appropriate length and color. This procedure is modified appropriately when the two rods are of different colors (opposite

[8]More precisely, a RaM can be emulated on and is polynomially equivalent to a RAM, and the latter can, in turn, be emulated on a Turing machine. Note that a unit (uniform) cost RAM model, or so-called UMRAM, is *not* polynomially equivalent to a Turing machine. Equivalence in both computational power and speed requires use of the logarithmic-cost RAM version. For further discussion, see Aho, Hopcroft, and Ullman [1974, Chapter 1], Kozen [1992, Sec. 1.3], and Traub and Werschulz [1998, Chapter 8].

[9]Recalling that a Turing machine tape is of potentially *infinite* length as is a word of a RaM, there is no difficulty, at least from a conceptual standpoint, in assuming a symbolic rod of *infinitesmal* width.

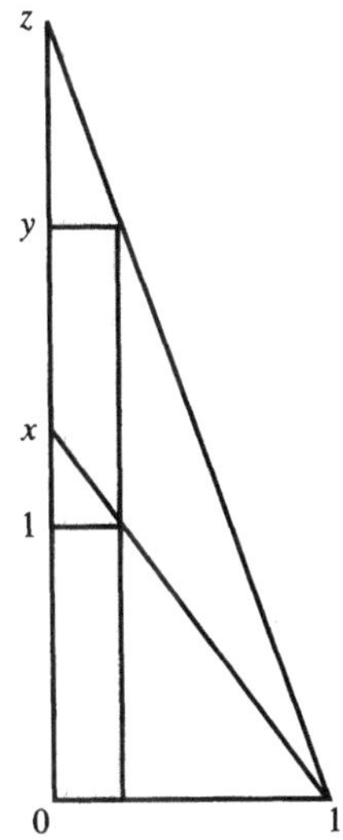

FIGURE 15.2 A GaM operation.

signs). The operation $\times$ between the geometric symbols representing two real numbers, say x and y, is performed by a geometric-scaling procedure, which is summarized in Figure 15.2 for the case where the two numbers are of the same sign (say positive). Let the geometric symbols or rods representing x and y correspond to the line segments $[0, x]$ and $[0, y]$, respectively. Similarity of triangles in the figure—the horizontal lines are parallel, as are the verticals—implies that the geometric symbol representing the product of the two numbers $z = xy$ corresponds to the line segment $[0, z]$; i.e., the latter is the result of the GaM operation $[0, x] \times [0, y]$. The procedure and generation of the resulting geometric symbol for the "product" can be performed in *a completely mechanical way by an idealized Euclidean geometer*; hence the acronym GaM chosen for the machine. The procedure requires simple modifications when the numbers are of opposite signs (symbols of different colors) and when the $\div$ operation is substituted for $\times$, and these are left as exercises for the reader. In each case, a symbolic rod of the appropriate length and color is returned by a geometric technique that can be performed mechanically. In other respects, the GaM is identical to a RaM and would be programmed in the same way.

We see that the GaM provides a computational platform for a *conceptual* algorithm, whereas a RaM provides a computational platform for an *implementable* algorithm (using terminology introduced in Chapters 3 and 5, and further considered in Section 15.4). These two models of computation represent *opposite sides of the universal watershed between the analog and the digital*, and the underlying issues are initimately related to those raised in the previous section to motivate the emerging discipline of algorithmic science.

Analog-computation models have been studied from the early days of computing. Nielsen and Chuang [2000] note the following:

> In the years since Turing, many different teams of researchers have noticed that certain types of analog computers can efficiently solve problems believed to have no efficient solution on a Turing machine.... Unfortunately for analog computation, it turns out that when realistic assumptions about the presence of noise in analog computers are made, their power disappears in all known instances; they cannot efficiently solve problems which are not efficiently solvable on a Turing machine. This lesson—that the effects of realistic noise must be taken into account in evaluating the efficiency of a computational model—was one of the great challenges....

The work on real-number machines of Blum et al. [1998]—a foundation for the study of conceptual algorithms—has restored this debate to center stage. At issue is the relevance of an *analog Turing machine model with an uncountably infinite symbolic vocabulary* vis-à-vis a digital Turing machine with a finite vocabulary.

15.3.3 Floating-Point Model/FRaM

In making the transition from conceptual/analog to implementable/digital, the *floating-point model* of Wilkinson [1963] has been of fundamental importance. The Wilkinson model falls again within the realm of the digital Turing machine and it is a particular version of a RaM, where a representable floating-point number is defined by a signed *t-digit* normalized *mantissa* in a number system with *base*, say, β, and by an *exponent*, say e. Each real number, say a, is *rounded* to a nearest representable rational number $\mathrm{fl}(a)$ defined by

$$\mathrm{fl}(a) = a(1+\delta), \quad |\delta| \le \tfrac{1}{2}\beta^{1-t}.$$

The representable numbers are rationals that are unevenly distributed over the real line; see Kahaner, Moler, and Nash [1989] for a depiction. Also, the exponent e of a number must fall within a specified range, say $-m \le e \le M$, where m and M are integers. Numbers with exponents outside of this range are said to spill (overflow or underflow). The four basic operations are assumed to satisfy axiomatic or idealized floating-point arithmetic, namely,

$$\mathrm{fl}(a * b) = (a * b)(1+\delta), \quad |\delta| \le \tfrac{1}{2}\beta^{1-t},$$

where a and b are representable numbers, $*$ denotes any of the four arithmetic operations $+, -, \times, \div$, and t and β are defined above. For more detail, see, for example, Nazareth [1987, Chapter 4].

Let us refer to a machine (or an implemented algorithm) based on the Wilkinson model by the acronym FRaM. The idea of a *stable* implemented algorithm, applied to given, not necessarily representable, input data, say, $[a_1, a_2, \ldots, a_n]$, is then captured by the following statement:

$$\text{FRaM}[\text{fl}(a_1), \ldots, \text{fl}(a_n)] = \{\text{GaM}[a_1(1+\epsilon_1), \ldots, a_n(1+\epsilon_n)]\}(1+\epsilon_{n+1}), \tag{15.1}$$

where the quantities $|\epsilon_i|, i = 1, \ldots, n+1$, are "small." In other words, the *computed* answer derived from representable initial data

$$\text{fl}(a_i) = (a_i)(1+\delta_i), \quad |\delta_i| \leq \tfrac{1}{2}\beta^{1-t}, \quad i = 1, \ldots, n,$$

is equal to a "small" perturbation of the result obtained from *exact* computation when the corresponding GaM is applied to "small" perturbations of the given initial data. The *true* answer is $\text{GaM}[a_1, \ldots, a_n]$. When the problem is *ill-conditioned*, the true answer can be quite different from the computed answer (15.1) obtained from the FRaM. This determination lies within the realm of perturbation theory.

The notions of a *stable* algorithm, the *condition number* of a problem, and an associated *perturbation analysis* are fundamental concepts of numerical mathematics, especially within the field of numerical linear algebra. Their extension more broadly to other problems addressed by algorithmic science, in particular, linear programming, is currently a topic of active research; see, for example, Renegar [1995b].

The subject of computability and complexity models that lies at the foundation of algorithmic science is a large one, and we have touched on it only very briefly here. For an in-depth discussion, see Blum, Cucker, Shub, and Smale [1998] and other works cited above.

15.4 Conceptual Algorithms

The distinction between a *conceptual* and an *implementable* algorithmic realization of a method was first emphasized by Polak [1971]. We have seen examples of conceptual algorithms in Chapters 3, 5, and 10, and they are naturally suited to the real-number models of computation of Section 15.3.2.

The key algorithmic idea underlying a conceptual algorithm or a family of such algorithms—see the wide variety itemized in Section 13.6—is usually rooted in a technique that is already well known but that is *approached from a fundamentally new angle, uses an innovative construct, or is applied in a previously unanticipated way.* For example, Karmarkar's breakthrough potential-reduction algorithm for linear programming was soon discovered to have a close connection to homotopy techniques for the numerical solution of systems of nonlinear equations and initial-value ODE problems, and to standard logarithmic-barrier techniques for nonlinear programming. Recognition of these deeper roots, in turn, led to fresh breakthroughs in the formulation of primal–dual interior-point algorithms for linear programming (Chapters 10 and 11). Potentials for linear programming served to motivate the potential-based Lagrangian globalization method for solving nonlinear equations (Chapter 3). The projective transformations of Karmarkar were found to be related to the collinear-scaling techniques of Davidon (Section 12.5). The model/metric interplay within Dikin's affine-scaling algorithm motivated a broader interpretation within the setting of nonlinear minimization (Chapters 2, 7, and 8). The QN relation and low-rank updating inspired the QC relation and diagonal updating (Chapter 9). And many other examples can be cited.

Conceptual families of algorithms often have well-defined *interrelationships and collectively exhibit an underlying mathematical structure.* We have discussed this, in detail, in Chapter 13 for the variable-metric and CG families of algorithms. In particular, a study of the intimate relationship between the BFGS variable-metric and nonlinear CG algorithms (conceptual versions that use exact line searches) set the stage for the subsequent discovery of important new classes of limited-memory and successive affine-reduced algorithms for large-scale unconstrained minimization. Again, as we have seen in Chapter 6, a reformulation of (univariate) golden-section direct search showed it to be closely related to the Nelder–Mead direct-search algorithm restricted to a single dimension. This relationship, in turn, suggested ways to enhance the multivariate NM algorithm and make it more amenable to theoretical analysis and possibly more efficient in practice, and it served to place the NM algorithm on a sounder (nonheuristic) theoretical footing. Again, many other algorithmic examples can be cited along these lines.

The study of underlying conceptual underpinnings and interrelationships and the detailed analysis of convergence, complexity, and numerical characteristics of conceptual algorithms, within the natural setting of real-number models, represent an essential first stage (Level 0) in the development of an optimization or equation-solving method.

The formulation of implementable versions within the appropriate rational or floating-point models of computation—RaM, FRaM, digital TM—comes next.

15.5 Implementable Algorithms

Having formulated a conceptual algorithm, it is useful to be able to *implement* it quickly in a powerful, operator-rich language, within the vernacular of computational mathematics. Such a language serves as a medium for communicating algorithmic ideas precisely. It permits coding and subsequent modification to be carried out with relative ease, even when this results in a sacrifice of efficiency with regard to computer time, computer storage, and numerical stability, and, in consequence, permits only artificial problems or only a subclass of realistic problems to be solved. Later it becomes necessary to attend to the practicalities of computation, namely, finite-precision arithmetic and the limitations of computer resources. Thus questions of numerical stability and efficiency must be addressed, and they usually imply substantial reformulation of data structures and calculations, leading to an implementation that is capable of solving problems more representative of those encountered in practice. Such an implementation provides more realistic evidence about algorithm performance and also addresses the research needs of the experienced user. (By experienced user we mean one familiar with both the problem being solved and the algorithm used to solve it.) Finally, we come to the everyday user who is not expected to be familiar with the details of the algorithm. Meeting his or her requirements requires another giant step in elaboration and refinement in order to obtain a production code or item of mathematical software that has a convenient user interface, is transportable, and solves problems automatically with a minimum of user intervention.

It is thus useful to think in terms of *a hierarchy of implementations* of a differentiable programming method, and convenient to distinguish three levels in the hierarchy corresponding to the initial, intermediate, and final stages of implementation as outlined above. These hierarchical levels can also be viewed as representing the stages of evolution from formulation and implementation of a *mathematical algorithm*, through reformulation and implementation of a viable *numerical algorithm*, to final implementation of good *mathematical software.*

15.5.1 Level 1: Mathematical

A mathematical algorithm is a very basic implementable version of a conceptual algorithm. At this root level, we remain within the domain of the algorithm inventor and computational mathematician, and we are again concerned with algorithmic theory that relates a relatively small number of individual mathematical algorithms to one another, thus providing a coherent overall structure, and with the formulation and analysis of these individual algorithms. Issues and analysis of global convergence, local rate of convergence, and complexity play a central role. At this level, one seeks to make the task of implementing a particular mathematical algorithm (or testing a conjecture about some property of an algorithm) as painless as possible, so as to obtain some initial computational experience. This experience often results in new insights and thus helps in laying out the basic features of an algorithm.

A powerful interactive language that uses the vernacular of computational mathematics provides a convenient "sketchpad" for working out algorithmic ideas or verifying conjectures on a computer. Ideally, such a language should provide a wide array of mathematical operators and data types and should also be easy to extend. The APL language, Iverson [1962], was a pioneering effort of this type. It can be used interactively, has convenient facilities for defining new operators, and makes the full facilities of the language available for investigating relationships between program variables at a breakpoint. A disadvantage is that the compact notation often makes programs difficult to read or decipher. When APL is used in interpretive mode, there is often a price to be paid in terms of program efficiency. More recent and very suitable languages for the purpose include Gauss, Matlab, Maple, and Mathematica. (For example, the conceptual procedures of Chapter 5 could be implemented very quickly in Matlab using its matrix operators and a unidimensional minimizer provided with its optimization toolbox.) Also, there is obviously no reason why a modern dialect of Fortran should not be used, particularly when it is enhanced by a collection of modules that makes available a wide range of mathematical operators. We shall refer to an experimental implementation developed in an extensible, operator-rich language as a level-1 implementation.

A level-1 routine is useful not only to give a precise definition to an algorithmic idea, but also to serve as a medium for communicating the idea. In particular, it provides a useful teaching tool. Often, the main reason for developing a level-1 implementation is experimentation, so rigorous standards of implementation need not be applied. A "quick and dirty" program involving unrealistic data management and a blasé attitude to potential numerical difficulties is often initially acceptable. Although one sometimes runs the risk of obtaining misleading computational information because of numerical error, a good algorithm should not be unduly victimized by finite-precision arithmetic. In addition to revealing conceptual difficulties, experimentation at this level can also reveal potential numerical difficulties that can then be addressed in developing a higher-level program. The aims are often geared to testing *viability* of an algorithm; i.e., the implementation may not be good enough to measure either efficiency or reliability, which are important testing criteria at subsequent levels.

A level-1 implementation will often incorporate just the *key features of an algorithm.* The problems that are addressed may be artificial or a subset of realistic problems, for example, smooth convex functions in nonlinear minimization. Convenience of problem specification may be emphasized. And often, a level-1 implementation will be designed simply to solve a *particular* problem, i.e., to "get the job done." Level-1 routines are generally programs with a short shelf life, which are intended to provide initial computational feedback or to furnish the solution of a particular problem, and they will then frequently be discarded, or modified beyond recognition.

15.5.2 Level 2: Numerical

A powerful computer language, no matter how convenient, does not circumvent (although it may postpone) the painstaking task involved in implementing a routine that is both numerically sound and efficient in its use of computer resources. At level 2, we are concerned with making a viable algorithmic idea workable within the setting of finite-precision arithmetic, the limitations of computer resources, and a problem space that is more representative of problems that arise in practice. We are now, so to speak, firmly within the province of the computational scientist and numerical analyst. For example, Dahlquist and Bjorck [1974] define a numerical algorithm as a complete description of well-defined operations whereby each permissible input data vector is transformed into an output data vector. By "operations" is meant the atomic and logical operations that a computer can perform, together with references to previously defined numerical algorithms. Round-off error analysis is a measure of the intellectual difficulties associated with algorithm reformulation and implementation at this level; see Wilkinson [1965]. Since a particular mathematical algorithm can be reformulated in a variety of ways, each with different numerical properties, we are in a larger "algorithmic space." For example, e^x can be computed by the series $\sum_{i=1}^{N}(x^i/i)$ for some suitable N. However, if x is negative, cancellation error can be avoided by reformulating this calculation as $1/\sum_{i=1}^{N}\{(-x)^i/i\}$. Algorithms at level 2 sometimes bear little resemblance to one another even when they are based on the same mathematical formulae, because the reformulation can involve a high degree of numerical detail.

Since one is concerned here with wider distribution of programs than at level 1 and with programs that have a longer "shelf life," it is more likely that a dialect of Fortran, C, Pascal, or similar language would be used (or pseudocode based on one of them). Since the variety of operators in these languages is quite limited, a collection of modules becomes especially useful here (modules can be thought of as a form of language extension). A modular collection also provides a "sketchpad" and a teaching tool at a more sophisticated level of implementation than level 1. Important criteria for both modules and level-2 implementations are *flexibility, ease of modification, generality, readability.* As already mentioned, attention should be paid both to numerical aspects and to data management, but relatively simple strategies (for example, very simple convergence criteria) may still he employed. The aim is to develop readable and modifiable implementations that can be employed by an experienced user to solve research problems and used for algorithm experimentation and development in a realistic problem setting. These two aims, of course, go well together. The goal of readability and modifiability for modules is especially important because it is unlikely that their design will be general enough for them to be used in simply a "plug-in" mode. Some changes will quite often have to be made to suit the particular implementation at hand.

As regards problem specification, we are concerned here with problems that are representative of those that arise in practice and for which realistic assumptions are made about availability of information. Testing would emphasize *efficiency* and *stability*, in order to investigate whether the algorithm is really effective. Effectiveness might be measured, for example, in terms of the number of function evaluations needed to find the optimal solution and the number of failures. There would also be the need for probing and tracing within the code as in level 1.

Good examples of level-2 implementations are the optimization routines available in the A.E.R.E. Harwell library, the nonlinear programming modules of Dennis and Schnabel [1983], or the linear programming modules in Nazareth [1986f].

15.5.3 Level 3: Quality Software

Finally, we come to the development of user-oriented production programs, or mathematical software. This falls within the province of the software engineer and requires yet another giant step in elaboration and refinement in order to arrive at a complete (perhaps even definitive) expression of an algorithm in a particular language (often Fortran) in a form capable of execution on a range of computers. It may incorporate complex adaptive strategies and tactics in order to further enhance efficiency. Usually, such software is expected to meet certain standards of quality, and at the very least it should be user-friendly and reliable. Just as the need for fine attention to detail in convergence, complexity, and error analysis typifies the inherent difficulties addressed at levels 1 and 2, here one might characterize the difficulties by a term such as software *systems analysis* or perhaps, more appropriately, *systems synthesis*. This also requires careful attention to detail, although this is less mathematical in nature, and careful coordination of many disparate components is necessary in order to synthesize from them a workable piece of software. Obviously, this task also requires an appreciation of the insights that convergence analysis and error analysis can provide and the principles that guide algorithm formulation, such as invariance under transformations of variables.

Fortran being the accepted language of scientific computation, it is the most common choice of implementation language. There is now an extensive literature on the development of high-quality mathematical software, premised on the pioneering work of Rice [1971] and Smith et al. [1974]. High-quality mathematical software is expected to be *reliable, robust, efficient, structured, well-documented, valid*, and *transportable.*

It is likely that the term "high-quality" or the phrase "reliable and robust" will be automatically assumed when we talk about mathematical software. (After all, no one talks nowadays about reliable and robust hardware. It is just expected to have these qualities.) It is also likely that the terms "package" and "library" will increasingly be associated only with software at this level. As noted, the software should be well structured and modular, but here "module" is used more in the sense of an efficient subroutine in a structured program, designed within the context of the overall implementation or package. It is much more likely that such modules would be used in a plug-in mode. Often, the design of level-3 software is too delicate for causal tampering, and sometimes such software is simply not designed to be modifiable.

Obviously, level-3 software should be able to solve problems as they arise in practice, and it is common for a package to include routines to check user-supplied information or to aid the process of specifying problems in a standard format. In the testing process increased emphasis is placed on reliability and robustness, and in the evaluation process routines are often treated as black boxes to be evaluated as one would any consumer product (see, for example, the performance profile approach of Lyness [1979]).

Good examples of level-3 optimization software can be found in the package Minpack-1 (Moré, Garbow, and Hillstrom [1980]) and in a variety of other software libraries currently available.

15.5.4 Interplay Between Levels

There is obviously nothing sacrosanct about identifying three levels in the hierarchy of implementation and obviously no clear-cut distinctions between them. Level-1 and -2 routines can and should be used to solve practical problems, and a level-3 implementation can be used to study the encoded algorithm and, by replacing part of the code, to develop and experiment with related algorthms. In general, however, one can say that while higher-level routines, in particular level-3 routines, are useful at a lower level, lower-level routines, in particular level 1, may not be of *direct* use at a higher level. However, a central theme of this discussion is that a lower-level implementation provides a *staging ground* for the development of an implementation at a higher level. In particular, *validation* of higher-level routines can be carried out using the results obtained from lower-level routines.

Another theme is that by drawing distinctions between levels of implementation one can usually clarify both the intent of a program and the meaning of terms like module, testing criteria, usability, and so on. They are usually determined by the level of implementation. For example, implementing a gradient method with line searches performed by bisection on the directional derivative can be done very quickly using Matlab. But a good level-3 implementation of steepest-descent would require careful craftsmanship and be much more time-consuming. For another example, consider the optimization "recipes" in Press, Vetterling, Teukolsky, and Flannery [1992]. Judged by level-1 criteria, they represent a valuable contribution, but the implementations are very crude when judged by the criteria of levels 2 and 3.

With the increasing availability of very powerful, operator-rich languages, and with scientists and engineers now being drawn increasingly into computational mathematics and algorithmics and making important contributions, it is likely that *level-1 implementation will become the center of action*, providing an arena for new algorithmic ideas and an environment for solving practical problems. The latter often have a great deal of special structure that must be taken into consideration in order to make them computationally tractable, and this characteristic makes them less amenable to solution by level-3, general purpose, software. To solve such problems often requires the development of tailor-made level-1 implementations. Levels 2 and 3 are more the province of skilled specialists. The level-3 implementations that they eventually develop then percolate back as new operators for use within powerful languages, for example, Matlab and Mathematica, at level 1; i.e., hierarchical implementation has a feedback loop that eventually enriches the level-1 environment.

15.6 Algorithmic Science

In conclusion, the emerging discipline discussed in previous sections of this chapter can be characterized as follows:

> *Algorithmic science is the theoretical and empirical study and the practical implementation and application of real-number algorithms, in particular for solving problems of continuous mathematics that arise within the fields of optimization and numerical analysis.*

Note that algorithms of discrete mathematics are not excluded in the above statement, but they take second place to algorithms of continuous mathematics.

An embracing algorithmic science discipline enhances the ability of researchers to recognize and take advantage of *interrelationships* of methods and associated algorithms to one another. It places at center stage the study of conceptual and implementable real-number algorithms as *objects in themselves* that have fascinating convergence, complexity, and numerical properties. It encourages a *cross-fertilization* of ideas, by bringing together people from various mathematical, scientific, and engineering areas of application who can approach the development of real-number algorithms from different perspectives, for example, essentialist vis-à-vis population-based. It emphasizes the *study of mathematical foundations* (see Section 15.3), which is fundamental to any scientific discipline. And it promotes the appropriate university (and other) *training* of new entrants to the field.

In this monograph we have been concerned with differentiable problems. When other areas are considered, the schematic given at the start of Part V of our monograph is seen to depict only a *central cross-section* of a much larger field embracing real-number algorithms for the solution of problems that may be nonsmooth, stochastic, infinite-dimensional, network-based, and so on: a rich and very beautiful scientific discipline with strong mathematical foundations and wide applicability.

15.7 Notes

Sections 15.1–15.4: This material is derived from an Algorithmic Science Research Center Feasibility Study (1999–2000) conducted with support from the Boeing Company.

Section 15.5: The discussion is derived from Nazareth [1985].

References

[1] Abramson, F. (1971), "Effective computation over the real numbers," in *12th Annual IEEE Symp. on Switching and Automata Theory*, 33–37.

[2] Adams, L. and Nazareth, J.L., eds. (1996), *Linear and Nonlinear Conjugate Gradient-Related Methods*, SIAM, Philadelphia.

[3] Adler, I., Karmarkar, N.K., Resende, M.C.G., and Veiga, G. (1989), "An implementation of Karmarkar's algorithm for linear programming," *Mathematical Programming*, 44, 297–335.

[4] Aho, A.V., Hopcroft, J.E., and Ullman, J.D. (1974), *The Design and Analysis of Computer Algorithms*, Addison-Wesley, Reading, Massachusetts.

[5] Al-Baali, M. (1985), "Descent property and global convergence of the Fletcher–Reeves method with inexact line searches," *IMA J. Numerical Analysis*, 5, 121–124.

[6] Allgower, E.L. (1998), private communication.

[7] Allgower, E.L. and Georg, K. (1990), *Numerical Continuation Methods: An Introduction*, Vol. 13, Series in Computational Mathematics, Springer-Verlag, Heidelberg, Germany.

[8] Andersen, E.D., Gondzio, J., Mészáros, C., and Xu, X. (1996), "Implementation of interior point methods for large scale linear program-

ming," in *Interior-Point Methods in Mathematical Programming*, T. Terlaky (ed.), Kluwer Academic Publishers, Dordrecht, 189–252.

[9] Anstreicher, K.M. (1996), "Potential-reduction algorithms," in *Interior-Point Methods in Mathematical Programming*, T. Terlaky (ed.), Kluwer Academic Publishers, Dordrecht, 125–158.

[10] Averick, B.M., Carter, R.G., Moré, J.J., and Xue, G. (1992), *The MINPACK-2 Test Problem Collection*, Preprint MCS-P153-0692, Mathematics and Computer Science Division, Argonne National Laboratory, Argonne, Illinois.

[11] Avis, D. and Chvatal, V. (1978), "Notes on Bland's pivoting rule," *Mathematical Programming Study*, 8, 24–34.

[12] Bäck, T. (1996), *Evolutionary Algorithms in Theory and Practice*, Oxford University Press, New York.

[13] Barnes, E.R. (1986), "A variation on Karmarkar's algorithm for linear programming," *Mathematical Programming*, 36, 174–182.

[14] Barnes, E.R., Chopra, S., and Jensen, D. (1987), "Polynomial-time convergence of the affine scaling algorithm with centering," presented at the conference *Progress in Mathematical Programming*, Asilomar Conference Center, Pacific Grove, California, March 1–4, 1987.

[15] Barrett, R., Berry, M., Chan, T.F., Demmel, J., Donato, J., Dongarra, J., Eijkhout, V., Pozo, R., Romine, C., and van der Vorst, H. (1993), *Templates for the Solution of Linear Systems*, SIAM, Philadelphia.

[16] Barzilai, J. and Borwein, J.M. (1988), "Two-point step size gradient methods," *IMA. J. Numerical Analysis*, 8, 141–148.

[17] Berlinski, D. (2000), *The Advent of the Algorithm: The Idea that Rules the World*, Harcourt, New York.

[18] Bertsekas, D.P. (1982), *Constrained Optimization and Lagrange Multiplier Methods*, Academic Press, New York.

[19] Bertsekas, D.P. (1999), *Nonlinear Programming*, Athena Scientific, Belmont, Massachusetts (second edition).

[20] Bland, R.G. (1977), "New finite pivoting rules for the simplex method," *Mathematics of Operations Research*, 2, 103–107.

[21] Blum, L., Cucker, F., Shub, M., and Smale S. (1998), *Complexity and Real Computation*, Springer-Verlag, New York (with a foreword by R.M. Karp).

[22] Bixby, R.E. (1994), "Progress in linear programming," *ORSA J. on Computing*, 6, 15–22.

[23] Bock, H.G., Kostina, E., and Schlöder, J.P. (2000), "On the role of natural level functions to achieve global convergence for damped Newton methods," in *System Modelling and Optimization: Methods, Theory and Applications*, M.J.D. Powell and S. Scholtes (eds.), Kluwer Academic Publishers, London, 51–74.

[24] Bounds, D.G. (1987), "New optimization methods from physics and biology," *Nature*, 329, 215–219.

[25] Brent, R.P. (1973), *Algorithms for Minimization without Derivatives*, Prentice-Hall, Englewood Cliffs, New Jersey.

[26] Broyden, C.G. (1970), "The convergence of a class of double-rank minimization algorithms," Parts I and II, *J.I.M.A.*, 6, 76–90, 222–236.

[27] Buckley, A. (1978), "Extending the relationship between the conjugate gradient and BFGS algorithms," *Mathematical Programming*, 15, 343–348.

[28] Burke, J.V. and Wiegmann, A. (1996), "Notes on limited-memory BFGS updating in a trust-region framework," Department of Mathematics, University of Washington, Seattle.

[29] Burke, J.V. and Xu, S. (2000), "A non-interior predictor–corrector path-following algorithm for the monotone linear complementarity problem," *Mathematical Programming*, 87, 113–130.

[30] Byrd, R.H., Nocedal, J., and Schnabel, R.B. (1994), "Representations of quasi-Newton matrices and their use in limited-memory methods," *Mathematical Programming*, 63, 129–156.

[31] CACR Annual Report (1997), Center for Advanced Computing Research, California Institute of Technology, Pasadena, California.

[32] Carter, R.G. (1993), "Numerical experience with a class of algorithms for nonlinear optimization using inexact function and gradient information," *SIAM J. Scientific Computing*, 14, 368–388.

[33] Cauchy. A. (1829), "Sur la détermination approximative des racines d'une équation algébrique ou transcendante," *Oeuvres Complete (II)*, 4, 573-607. Gauthier-Villars, Paris, 1899.

[34] Cauchy, A. (1847), "Analyse mathématique-méthode générale pour la résolution des systèmes d'équations simultanées," *Comp. Rend. Acad. Sci.*, Paris, 536–538.

[35] Chalmers, A. and Tidmus, J. (1996), *Practical Parallel Processing: An Introduction to Problem Solving in Parallel*, International Thomson Computer Press, London, England.

[36] Chen, B. and Harker, P. (1993), "A non-interior-point continuation method for linear complementarity problems," *SIAM J. Matrix Anal. Appl.*, 14, 1168–1190.

[37] Chow, S.N., Mallet-Paret, J., and Yorke, J.A. (1978), "Finding zeros of maps: homotopy methods that are constructive with probability one," *Mathematics of Computation*, 32, 887–899.

[38] Chvatal, V. (1983), *Linear Programming*, W.H. Freeman and Co., New York.

[39] Conn, A.R., Gould, N.I.M., and Toint Ph.L. (1991), "Convergence of quasi-Newton matrices generated by the symmetric rank one update," *Mathematical Programming*, 50, 177–196.

[40] CPLEX User's Guide (1993), CPLEX Optimization, Incline Village, Nevada.

[41] Dahlquist, G. and Bjorck, A. (1974), *Numerical Methods*, Prentice-Hall, Englewood Cliffs, New Jersey.

[42] Dai, Y.H. (1997), *Analyses of Nonlinear Conjugate Gradient Methods*, Ph.D. dissertation, Institute of Computational Mathematics and Scientific and Engineering Computing, Chinese Academy of Sciences, Beijing, China.

[43] Dai, Y.H. and Yuan, Y. (1996), "Convergence properties of the Fletcher–Reeves method," *IMA J. Numer. Anal.*, 16, 155–164.

[44] Dai, Y.H. and Yuan, Y. (1998), "A class of globally convergent conjugate gradient methods," Research Report No. ICM-98-30, Institute of Computational Mathematics and Scientific and Engineering Computing, Chinese Academy of Sciences, Beijing, China.

[45] Dai, Y.H. and Yuan, Y. (1999), "A nonlinear conjugate gradient method with a strong global convergence property," *SIAM Journal on Optimization*, 10, 177–182.

[46] Dantzig, G.B. (1963), *Linear Programming and Extensions*, Princeton University Press, Princeton, New Jersey.

[47] Dantzig, G.B. (1980), "Expected number of steps of the simplex method for a linear program with a convexity constraint," Technical Report SOL 80-3, Systems Optimization Laboratory, Department of Operations Research, Stanford University, Stanford, California.

[48] Dantzig, G.B. (1983), "Reminiscences about the origins of linear programming," in *Mathematical Programming: The State of the Art, Bonn, 1982*, A. Bachem, M. Grotschel and B. Korte (eds.), Springer-Verlag, Berlin, 78–86.

[49] Dantzig, G.B. (1985), "Impact of linear programming on computer development," Technical Report SOL 85-7, Systems Optimization Laboratory, Department of Operations Research, Stanford University, Stanford, California.

[50] Dantzig, G.B. and Thapa, M.N. (1997), *Linear Programming: I: Foundations*, Springer Series in Operations Research, Springer-Verlag, New York.

[51] Davidon, W.C. (1959), "Variable metric method for minimization," Argonne National Laboratory, Report ANL-5990 (Rev.), Argonne, Illinois (reprinted, with a new preface, in *SIAM J. Optimization*, 1, 1–17, 1991).

[52] Davidon, W.C. (1980), "Conic approximations and collinear scalings for optimizers," *SIAM J. Numer. Anal.*, 17, 268–281.

[53] Davidon, W.C., Mifflin, R.B., and Nazareth, J.L. (1991), "Some comments on notation for quasi-Newton methods," *Optima*, 32, 3–4.

[54] Dennis, J.E. and Moré, J.J. (1977), "Quasi-Newton methods: motivation and theory," *SIAM Review*, 19, 46–89.

[55] Dennis, J.E. and Schnabel, R.B. (1983), *Numerical Methods for Unconstrained Optimization and Nonlinear Equations*, Prentice-Hall, New Jersey.

[56] Dennis, J.E. and Torcson, V. (1991), "Direct search methods on parallel machines," *SIAM J. Optimization*, 1, 448–474.

[57] Dennis, J.E. and Wolkowicz, H. (1993), "Sizing and least change secant updates," *SIAM J. Numerical Analysis*, 10, 1291–1314.

[58] Dikin, I.I. (1967), "Iterative solution of problems of linear and quadratic programming," *Soviet Mathematics Doklady*, 8, 674–675.

[59] Dikin, I.I. (1974), "On the convergence of an iterative process," *Upravlyaemye Sistemi.* 12, 54–60.

[60] Ding, J. and Li, T. (1990), "An algorithm based on weighted logarithmic barrier functions for linear complementarity problems," *Arabian J. Sciences and Engineering*, 15, 4(B), 679–685.

[61] Ding, J. and Li, T. (1991), "A polynomial-time predictor–corrector algorithm for a class of linear complementarity problems," *SIAM J. Optimization*, 1, 83–92.

[62] Di Pillo, G. and Grippo, L. (1979), "A new class of augmented Lagrangians in nonlinear programming," *SIAM Journal on Control and Optimization*, 17, 618–628.

[63] Dixon, L.C.W. and Nazareth, J.L. (1996), "Potential functions for non-symmetric sets of linear equations," presented at *SIAM Conference on Optimization*, May 20–22, Victoria, B.C., Canada.

[64] Edelman, G.E. (1988), *Topobiology*, Basic Books, New York.

[65] Edelman, G.E. (1992), *Bright Air, Brilliant Fire: On the Matter of the Mind*, Basic Books, New York.

[66] El-Bakry, A.S., Tapia, R.A., Zhang, Y., and Tsuchiya, T. (1992), "On the formulation and theory of the Newton interior-point method for nonlinear programming," Report TR 92-40, Department of Computational and Applied Mathematics, Rice University, Houston, Texas (revised April, 1995).

[67] Ermoliev Y. and Gaivoronski, A. (1994), "Stochastic quasigradient methods," *SIAG/OPT Views-and-News*, 4, 7–10.

[68] Ermoliev, Y. and Wets, R.J.-B., eds., (1988), *Numerical Techniques for Stochastic Optimization Problems*, Springer-Verlag, Berlin, Germany.

[69] Fiacco, A.V. and McCormick, G.P. (1968), *Nonlinear Programming: Sequential Unconstrained Minimization Techniques*, Wiley, New York.

[70] Fletcher, R. (1970), "A new approach to variable metric algorithms," *Computer Journal*, 13, 317–322.

[71] Fletcher, R. (1980), *Practical Methods of Optimization: Volume 1*, Wiley, New York.

[72] Fletcher, R. and Powell, M.J.D. (1963), "A rapidly convergent descent method for minimization," *Computer Journal*, 6, 163–168.

[73] Fletcher, R. and Reeves, C. (1964), "Function minimization by conjugate gradients," *Computer Journal*, 6, 149–154.

[74] Floudas, C.A. and Pardalos, P.M., eds. (2001), *Encyclopedia of Optimization*, Vols. I–VI, Kluwer Academic Publishers, Dordrecht and Boston.

[75] Fogel, D.B. (1992), "Evolutionary optimization," ORINCON Corporation, San Diego, California (preprint).

[76] Fogel, D.B. (1995), *Evolutionary Computation: Toward a New Philosophy of Machine Intelligence*, IEEE Press, Piscataway, New Jersey.

[77] Fogel, L.J., Owens, A.J., and Walsh, M.J. (1966), *Artificial Intelligence through Simulated Evolution*, Wiley, New York.

[78] Fox, B.L. (1993), "Integrating and accelerating tabu search, simulated annealing and genetic algorithms," *Annals of Operations Research*, 41, 47–67.

[79] Fox, B.L. (1995), "Uprooting simulated annealing and genetic algorithms," *SIAG/OPT Views-and-News*, 7, 5–9.

[80] Fox, G.C., Williams, R.D., and Messina, P.C. (1994), *Parallel Computing Works*, Morgan Kauffman Publishers, San Francisco, California.

[81] Freund, R.M. (1996), "An infeasible-start algorithm for linear programming whose complexity depends on the distance from the starting point to the optimal solution," *Annals of Operations Research*, 62, 29–58.

[82] Frisch, K.R. (1955), "The logarithmic potential method for convex programming," manuscript, Institute of Economics, University of Oslo, Oslo, Norway.

[83] Frisch, K.R. (1956), "La résolution des problèmes de programme linéaire par la méthode du potential logarithmique," *Cahiers du Seminaire D'Económetrie*, 4, 7–20.

[84] Garcia, C.B. and Zangwill, W.I. (1981), *Pathways to Solutions, Fixed Points and Equilibria*, Prentice-Hall, Englewood Cliffs, New Jersey.

[85] Garey, M. and Johnson, D. (1979), *Computers and Intractability: A Guide to the Theory of NP-Completeness*, Freeman, New York.

[86] Gay, D.M. (1985), "Electronic mail distribution of linear programming test problems," *Mathematical Programming Society COAL Newsletter*, 13, 10–12.

[87] Gilbert, J.C. and Lemaréchal, C. (1989), "Some numerical experiments with variable-storage quasi-Newton algorithms," *Mathematical Programming*, Series B, 45, 407–435.

[88] Gill, P.E. and Murray, W. (1979), "Conjugate gradient methods for large-scale nonlinear optimization," Technical Report SOL 79-15, Department of Operations Research, Stanford University, Stanford, California.

[89] Gill, P.E., Murray, W., Ponceleón, D.B., and Saunders, M.A. (1991), "Primal-dual methods for linear programming," Technical Report SOL 91-3, Systems Optimization Laboratory, Stanford University, Stanford, California.

[90] Gill, P.E., Murray, W., Saunders, M.A., Tomlin, J.A., and Wright, M.H. (1986), "On projected Newton barrier methods for linear programming and an equivalence to Karmarkar's projective method," *Mathematical Programming*, 36, 183–209.

[91] Goldman, A.J. and Tucker, A.W. (1956), "Theory of linear programming," in *Linear Inequalities and Related Systems*, H.W. Kuhn and A.W. Tucker (eds.), Annals of Mathematical Studies, 38, Princeton University Press, Princeton, New Jersey, 53–97.

[92] Goldstein, A.A. (1967), *Constructive Real Analysis*, Harper's Series in Modern Mathematics, Harper and Row, New York.

[93] Goldstein, A.A. (1991), "A global Newton method," DIMACS, Vol. 4, *Applied Geometry and Discrete Mathematics*, American Mathematical Society, Providence, Rhode Island, 301–307.

[94] Golub, G.H. and O'Leary, D.P. (1989), "Some history of the conjugate gradient and Lanczos algorithms: 1948–1976," *SIAM Review*, 31, 50–102.

[95] Golub, G.H. and Van Loan, C.F. (1983), *Matrix Computations*, Johns Hopkins University Press, Baltimore, Maryland. (second edition, 1989.)

[96] Gondzio, J. (1995), "HOPDM (version 2.12)—a fast LP solver based on a primal–dual interior point method," *European J. Oper. Res.*, 85, 221–225.

[97] Gonzaga, C.C. (1989), "Conical projection methods for linear programming," *Mathematical Programming*, 43, 151–173.

[98] Gonzaga, C.C. (1992), "Path-following methods for linear programming," *SIAM Review*, 34, 167–227.

[99] Gould, S.J. (1996), *Full House: The Spread of Excellence from Plato to Darwin*, Harmony Books, Crown Publishers, New York.

[100] Graham, R., Knuth, D.E., and Patashnik, O. (1989), *Concrete Mathematics: A Foundation for Computer Science*, Addison-Wesley, Reading, Massachusetts.

[101] Griewank, A. (1989), "On automatic differentiation," in *Mathematical Programming: Recent Developments and Applications*, M. Iri and K. Tanabe (eds.), Kluwer Academic Publishers, Dordrecht, 83–108.

[102] Gropp, W., Lusk, E., and Skjellum, A. (1994), *Using MPI: Portable Parallel Programming with the Message-Passing Interface*, The MIT Press, Cambridge, Massachusetts.

[103] Guo, J. (1993), *A Class of Variable Metric-Secant Algorithms for Unconstrained Minimization*, Ph.D. dissertation, Department of Pure and Applied Mathematics, Washington State University, Pullman, Washington.

[104] Hamming, R. (1977), "Confessions of a numerical analyst," Division of Mathematics and Statistics (DMS) Newsletter, No. 33, CSIRO, Glen Osmond, South Australia.

[105] Hestenes, M.R. (1969), "Multiplier and gradient methods," *J.O.T.A.*, 4, 303–320.

[106] Hestenes, M.R. (1980), *Conjugate Direction Methods in Optimization*, Applications of Mathematics Series, 12, Springer-Verlag, Heidelberg and New York.

[107] Hestenes, M.R. and Stiefel, E.L. (1952), "Methods of conjugate gradients for solving linear systems," *J. Res. Nat. Bur. Stds.*, Section B, 49, 409–436.

[108] Hillstrom, K.E. (1976a), "Optimization routines in AMDLIB," ANL-AMD Tech. Memo. 297, Applied Mathematics Division, Argonne National Laboratory, Argonne, Illinois.

[109] Hillstrom, K.E. (1976b), "A simulation test approach to the evaluation and comparison of unconstrained nonlinear optimization algorithms," Report ANL-76-20, Argonne National Laboratory, Argonne, Illinois.

[110] Hirsch, M.W. and Smale, S. (1979), "On algorithms for solving $F(x) = 0$," *Comm. Pure Appl. Math.*, 32, 281–312.

[111] Holland, J.H. (1975), *Adaptation in Natural and Artificial Systems*, The University of Michigan Press, Ann Arbor, Michigan.

[112] Holland, J.H. (1995), *Hidden Order*, Addison-Wesley, Reading, Massachusetts.

[113] Hopcroft, J.E. and Kennedy, K.W., Chairs (1989), *Computer Science: Achievements and Opportunities*, Report of the NSF Advisory Committee for Computer Research, SIAM Reports on Issues in the Mathematical Sciences, SIAM, Philadelphia.

[114] Huard, P. (1967), "Resolution of mathematical programming with nonlinear constraints by the method of centers," in *Nonlinear Programming*, J. Abadie (ed.), North Holland, Amsterdam, 207–219.

[115] Iverson, K.E. (1962), *A Programming Language*, Wiley, New York.

[116] Jansen, B., Roos, C., and Terlaky, T. (1993), "A polynomial primal–dual Dikin-type algorithm for linear programming," Report No. 93-36, Faculty of Technical Mathematics and Informatics, Delft University of Technology, the Netherlands.

[117] Jansen, B., Roos, C., and Terlaky, T. (1996), "Interior point methods a decade after Karmarkar—a survey, with application to the smallest eigenvalue problem," *Statistica Neerlandica*, 50, 146–170.

[118] Kahaner, D., Moler, C., and Nash, S. (1989), *Numerical Methods and Software*, Prentice-Hall, Englewood Cliffs, New Jersey.

[119] Kantorovich, L.V. (1939), "Mathematical methods in the organization and planning of production," Publication House of the Leningrad State University. Translated in *Management Science*, 6, 1960, 366–422.

[120] Karmarkar, N. (1984), "A new polynomial-time algorithm for linear programming," *Combinatorica*, 4, 373–395.

[121] Karmarkar, N. (1995), presentation at the conference *Mathematics of Numerical Analysis: Real Number Algorithms*, August, 1995, Park City, Utah.

[122] Kauffman, S.A. (1993), *The Origins of Order: Self-Organization and Selection in Evolution*, Oxford University Press, Oxford and New York.

[123] Keller, H.B. (1978), "Global homotopies and Newton methods," in *Recent Advances in Numerical Analysis*, C. de Boor and G.H. Golub (eds.), Academic Press, New York, 73–94.

[124] Keller, H.B. (1987), *Lectures on Numerical Methods in Bifurcation Problems*, Springer-Verlag, Heidleberg, Germany.

[125] Kelley, C.T. (1999), *Iterative Methods for Optimization*, SIAM, Philadelphia.

[126] Kim, K. and Nazareth, J.L. (1994), "A primal null-space affine scaling method," *ACM Transactions on Mathematical Software*, 20, 373–392.

[127] Knuth, D.E. (1968), *The Art of Computer Programming*, Addison-Wesley, Reading, Massachusetts.

[128] Knuth, D.E. (1996), *Selected Papers on Computer Science*, CSLI Lecture Notes No. 59, Cambridge University Press, Cambridge, England.

[129] Kolda, T.G., O'Leary, D.P., and Nazareth, J.L. (1998), "BFGS with update skipping and varying memory," *SIAM J. on Optimization*, 8, 1060–1083.

[130] Kojima, M., Megiddo, N., and Mizuno, S. (1993), "A primal–dual infeasible interior point algorithm for linear programming," *Mathematical Programming*, 61, 263–280.

[131] Kojima, M., Megiddo, N., and Noma, T. (1991), "Homotopy continuation methods for nonlinear complementarity problems," *Mathematics of Operations Research*, 16, 754–774.

[132] Kojima, M., Megiddo, N., Noma, T., and Yoshise, A. (1991), *A Unified Approach to Interior Point Algorithms for Linear Complementarity Problems*, Lecture Notes in Computer Science 538, Springer-Verlag, Germany.

[133] Kojima, M., Mizuno, S., and Yoshise, A. (1989), "A primal–dual interior point algorithm for linear programming," in *Progress in Mathematial Programming: Interior-Point and Related Methods*, N. Megiddo (ed.), Springer-Verlag, New York, 29–47.

[134] Kojima, M., Mizuno, S., and Yoshise, A. (1991), "An $O(\sqrt{n}L)$ iteration potential reduction algorithm for linear complementarity problems," *Mathematical Programming*, 50, 331–342.

[135] Kozen, D.C. (1992), *The Design and Analysis of Algorithms*, Texts and Monographs in Computer Science, Springer-Verlag, New York.

[136] Kranich, E. (1991), "Interior point methods in mathematical programming: a bibliography," Discussion Paper 171, Institute of Economy and Operations Research, Fern Universität Hagen, P.O. Box 940, D-5800 Hagen 1, West Germany. Available through NETLIB: Send e-mail to netlib@research.att.com.

[137] Kuhn, H. and Quandt, R.E. (1963), "An experimental study of the simplex method," in *Experimental Arithmetic, High-Speed Computing and Mathematics*, N.C. Metropolis et al. (eds.), Proceedings of Symposium on Applied Mathematics XV, American Mathematical Society, Providence, Rhode Island, 107–124.

[138] Kuznetsov, A.G. (1992), "Nonlinear optimization toolbox," Report OUEL 1936/92, Department of Engineering Science, University of Oxford, Oxford, United Kingdom.

[139] Lagarias, J.C. (1993), "A collinear scaling interpretation of Karmarkar's linear programming algorithm," *SIAM J. on Optimization*, 3, 630–636.

[140] Lagarias, J.C., Reeds, J.A., Wright, M.H., and Wright, P.E. (1998) "Convergence properties of the Nelder–Mead simplex algorithm in low dimensions," *SIAM J. on Optimization*, 9, 112–147.

[141] Leonard, M.W. (1995), *Reduced Hessian Quasi-Newton Methods for Optimization*, Ph.D. dissertation, University of California, San Diego, California.

[142] Levine, D. (1996), "Users Guide to the PGAPack Parallel Genetic Algorithm Library," Report ANL-95/18, Mathematics and Computer Science Division, Argonne National Laboratory, Argonne, Illinois.

[143] Liu, D.C. and Nocedal, J. (1989), "On the limited memory method for large scale optimization," *Mathematical Programming*, Series B, 45, 503–528.

[144] Luenberger, D.G. (1984), *Linear and Nonlinear Programming*, Addison-Wesley (second edition).

[145] Lustig, I.J. (1991), "Feasibility issues in a primal–dual interior method for linear programming," *Mathematical Programming*, 49, 145–162.

[146] Lustig, I.J., Marsten, R.E., and Shanno, D. (1992), "On implementing Mehrotra's predictor–corrector interior point method for linear programming," *SIAM J. on Optimization*, 2, 435–449.

[147] Lustig, I.J., Marsten, R.E., and Shanno, D. (1994a), "Computational experience with a globally convergent primal–dual predictor–corrector interior algorithm for linear programming," *Mathematical Programming*, 66, 123–135.

[148] Lustig, I.J., Marsten, R.E., and Shanno, D. (1994b), "Interior point methods: computational state of the art," *ORSA J. on Computing*, 6, 1–15.

[149] Lyness, J.N. (1979), "A benchmark experiment for minimization algorithms," *Mathematics of Computation*, 33, 249–264.

[150] Mangasarian, O.L. (1969), *Nonlinear Programming*, McGraw-Hill, New York.

[151] Mangasarian, O.L. (1975), "Unconstrained Lagrangians in nonlinear programming," *SIAM Journal on Control*, 13, 772–791.

[152] Mayr, E. (1982), *The Growth of Biological Thought: Diversity, Evolution and Inheritance*, Harvard University Press, Cambridge, Massachusetts.

[153] Mayr, E. (2001), *What Evolution Is*, Basic Books, New York.

[154] McLinden, L. (1980), "The analogue of Moreau's proximation theorem, with applications to the nonlinear complementarity problem," *Pacific Journal of Mathematics*, 88, 101–161.

[155] McKinnon, K.I.M., (1998), "Convergence of the Nelder–Mead simplex method to a non-stationary point," *SIAM J. on Optimization*, 9, 148–158.

[156] Megiddo, N. (1989), "Pathways to the optimal set in linear programming," in *Progress in Mathematial Programming: Interior-Point and Related Methods*, N. Megiddo (ed.), Springer-Verlag, New York, 131–158.

[157] Megiddo, N. (1991), "On finding primal- and dual-optimal bases," *ORSA J. on Computing*, 3, 63–65.

[158] Mehrotra, S. (1991), "Finding a vertex solution using an interior point method," *Linear Algebra and Applications*, 152, 233–253.

[159] Mehrotra, S. (1992), "On the implementation of a (primal–dual) interior point method," *SIAM J. on Optimization*, 2, 575–601.

[160] Messina, P. (1997), "High-performance computers: the next generation," Parts I and II, *Computers in Physics*, 11, Nos. 5 and 6, 454–466, 599–610.

[161] Michalewicz, Z. (1996), *Genetic Algorithms + Data Structures = Evolution Programs*, (third, revised and extended edition), Springer-Verlag, Berlin, Germany.

[162] Mifflin, R.B. (1994), "A superlinearly convergent algorithm for minimization without evaluating derivatives," *Mathematical Programming*, 9, 100–117.

[163] Mifflin, R.B. and Nazareth, J.L. (1994), "The least prior deviation quasi-Newton update," *Mathematical Programming*, 65, 247–261.

[164] Mizuno, S. (1994), "Polynomiality of infeasible-interior-point algorithms for linear programming," *Mathematical Programming*, 67, 109–119.

[165] Mizuno, S. and Jarre, F. (1996), "Global and polynomial-time convergence of an infeasible-interior-point algorithm using inexact computation," Research Memorandum No. 605, The Institute of Statistical Mathematics, Tokyo, Japan.

[166] Mizuno, S., Kojima, M., and Todd, M.J. (1995), "Infeasible-interior-point primal–dual potential-reduction algorithms for linear programming," *SIAM J. on Optimization*, 5, 52–67.

[167] Mizuno, S., Todd, M.J., and Ye, Y. (1993), "On adaptive step primal–dual interior-point algorithms for linear programming," *Mathematics of Operations Research*, 18, 964–981.

[168] Mizuno, S., Todd, M.J., and Ye, Y. (1995), "A surface of analytic centers and infeasible-interior-point algorithms for linear programming," *Mathematics of Operations Research*, 20, 52–67.

[169] Monteiro, R.D.C, Adler, I., and Resende, M.G.C. (1990), "A polynomial-time primal–dual affine scaling algorithm for linear and convex quadratic programming and its power series extension," *Mathematics of Operations Research*, 15, 191–214.

[170] Moré, J.J. (1983), "Recent developments in algorithms and software for trust region methods," in *Mathematical Programming: The State of the Art, Bonn, 1982*, A. Bachem, M. Grotschel and B. Korte (eds.), Springer-Verlag, Berlin, 258–287.

[171] Moré, J.J. (1993), "Generalizations of the trust region problem," *Optimization Methods and Software*, 2, 189–209.

[172] Moré, J.J., Garbow, B.S., and Hillstrom, K.E. (1980), "User's Guide to Minpack-1," Report ANL-80-74, Argonne National Laboratory, Argonne, Illinois.

[173] Moré, J.J., Garbow, B.S., and Hillstrom, K.E. (1981), "Testing unconstrained optimization software," *ACM Transactions on Mathematical Software*, 7, 17–41.

[174] Moré, J.J. and Thuente. D.J. (1994), "Line search algorithms with guaranteed sufficient decrease," *ACM Transactions on Mathematical Software*, 20, 286–307.

[175] Nazareth, J.L. (1976), "A relationship between the BFGS and conjugate gradient algorithms," Tech. Memo. ANL-AMD 282, Applied Mathematics Division, Argonne National Laboratory, Argonne, Illinois. (Presented at the SIAM-SIGNUM Fall 1975 Meeting, San Francisco, California.)

[176] Nazareth, J.L. (1977), "MINKIT—an optimization system," Tech. Memo. ANL-AMD 305, Applied Mathematics Division, Argonne National Laboratory, Argonne, Illinois.

[177] Nazareth, J.L. (1979), "A relationship between the BFGS and conjugate gradient algorithms and its implications for new algorithms," *SIAM J. on Numerical Analysis*, 16, 794–800.

[178] Nazareth, J.L. (1984), "An alternative variational principle for variable metric updating," *Mathematical Programming*, 30, 99–104.

[179] Nazareth, J.L. (1985), "Hierarchical implementation of optimization methods," in *Numerical Optimization, 1984*, P. Boggs, R.H. Byrd, and R.B. Schnabel (eds.), SIAM, Philadelphia, 199–210.

[180] Nazareth, J.L. (1986a), "Conjugate gradient algorithms less dependent on conjugacy," *SIAM Review*, 28, 501–511.

[181] Nazareth, J.L. (1986b), "The method of successive affine reduction for nonlinear minimization," *Mathematical Programming*, 35, 97–109.

[182] Nazareth, J.L. (1986c), "An algorithm based on successive affine reduction and Newton's method," in *Proceedings of the Seventh INRIA International Conference on Computing Methods in Applied Science and Engineering*, (Versailles, France), R. Glowinski and J.-L. Lions (eds.), North-Holland, 641–646.

[183] Nazareth, J.L. (1986d), "Homotopy techniques in linear programming," *Algorithmica*, 1, 529–535.

[184] Nazareth, J.L. (1986e), "Analogues of Dixon's and Powell's theorems for unconstrained minimization with inexact line searches," *SIAM J. on Numerical Analysis*, 23, 170–177.

[185] Nazareth, J.L. (1986f), "Implementation aids for optimization algorithms that solve sequences of linear programs," *ACM Transactions on Mathematical Software*, 12, 307–323.

[186] Nazareth, J.L. (1987), *Computer Solution of Linear Programs*, Oxford University Press, Oxford and New York.

[187] Nazareth, J.L. (1989), "Pricing criteria in linear programming," in *Progress in Mathematical Programming*, N. Megiddo (ed.), Springer-Verlag, New York, 105–129.

[188] Nazareth, J.L. (1990), "Some parallels in the historical development of the simplex and interior point algorithms for linear programming," WSU Mathematics Notes, Vol 33, No. 1, Washington State University, Pullman, Washington.

[189] Nazareth, J.L. (1991), "The homotopy principle and algorithms for linear programming," *SIAM J. on Optimization*, 1, 316–332.

[190] Nazareth, J.L. (1993a), "Soul-searching in the mathematical, computer and decision sciences," Editorial Feature, *SIAG/OPT Views-and-News*, No. 2, 1–2.

[191] Nazareth, J.L. (1993b), "From Dikin to Karmarkar via scalings, models and trust regions," *SIAG/OPT Views-and-News*, No. 2, 6–9.

[192] Nazareth, J.L. (1994a), "The Newton and Cauchy perspectives on computational nonlinear optimization," *SIAM Review*, 36, 215–225.

[193] Nazareth, J.L. (1994b), *The Newton–Cauchy Framework: A Unified Approach to Unconstrained Nonlinear Minimization*, Lecture Notes in Computer Science Series, Vol. 769, Springer-Verlag, Berlin and New York.

[194] Nazareth, J.L. (1994c), "Quadratic and conic approximating models in linear programming," *Mathematical Programming Society COAL Bulletin*, Vol. 23.

[195] Nazareth, J.L. (1994d), "A reformulation of the central path equations and its algorithmic implications," Technical Report 94-1, Department of Pure and Applied Mathematics, Washington State University, Pullman, Washington.

[196] Nazareth, J.L. (1995a), "A framework for interior methods of linear programming," *Optimization Methods and Software*, 5, 227–234.

[197] Nazareth, J.L. (1995b), "If quasi-Newton then why not quasi-Cauchy?," *SIAG/OPT Views-and-News*, No. 6, 11–14.

[198] Nazareth, J.L. (1995c), "The quasi-Cauchy method: a stepping stone to derivative-free algorithms," Technical Report 95-3, Department of Pure and Applied Mathematics, Washington State University, Pullman, Washington.

[199] Nazareth, J.L. (1995d), "Trust regions based on conic functions in linear and nonlinear programming," *Numerical Linear Algebra with Applications*, 2, 235–241.

[200] Nazareth, J.L. (1996a), "Lagrangian globalization: solving nonlinear equations via constrained optimization," in *Mathematics of Numerical Analysis: Real Number Algorithms*, J. Renegar, M. Shub, and S. Smale (eds.), American Mathematical Society, Providence, Rhode Island, 533–542.

[201] Nazareth, J.L. (1996b), "The implementation of linear programming algorithms based on homotopies," *Algorithmica*, 15, 332–350.

[202] Nazareth, J.L. (1996c), "A view of conjugate gradient-related algorithms for nonlinear optimization," in *Linear and Nonlinear Conjugate Gradient-Related Methods*, L. Adams and J.L. Nazareth (eds.), SIAM, Philadelphia, 149–164.

[203] Nazareth, J.L. (1997a), "Deriving potential functions via a symmetry principle for nonlinear equations," *Operations Research Letters*, 21, 147–152.

[204] Nazareth, J.L. (1997b), "Multialgorithms for parallel computing." Invited presentation at the *Pacific Northwest Numerical Analysis Seminar*, The Boeing Company, Seattle, Washington (October 18, 1997).

[205] Nazareth, J.L. (1998), "Computer solution of linear programs: non-simplex algorithms," in *Advances in Nonlinear Programming*, Y. Yuan (ed.), Kluwer Academic Publishers, Dordrecht and Boston, 119–151.

[206] Nazareth, J.L. (2000), Book Review: *Iterative Methods for Optimization* by C.T. Kelley (SIAM, Philadelphia), *SIAM Review*, 42, 535–539.

[207] Nazareth, J.L. (2001a), "Unconstrained nonlinear optimization: Newton–Cauchy framework," in *Encyclopedia of Optimization*, C.A. Floudas and P.M. Pardalos (eds.), Kluwer Academic Publishers, Dordrecht and Boston, Vol. V (R-Z), 481–486.

[208] Nazareth, J.L. (2001b), "Conjugate gradient methods," in *Encyclopedia of Optimization*, C.A. Floudas and P.M. Pardalos (eds.), Kluwer Academic Publishers, Dordrecht and Boston, Vol. I (A–D), 319–323.

[209] Nazareth, J.L. (2001c), "Multialgorithms for parallel computing: a new paradigm for optimization," in *Stochastic Optimization: Algorithms and Applications*, S. Uryasev and P.M. Pardalos (eds.), Kluwer Academic Publishers, Dordrecht and Boston, 183–222.

[210] Nazareth, J.L. and Ariyawansa, K.A. (1989), "On accelerating Newton's method based on a conic model," *Information Processing Letters*, 30, 277–281.

[211] Nazareth, J.L. and Qi, L. (1996), "Globalization of Newton's method for solving nonlinear equations," *Numerical Linear Algebra with Applications*, 3, 239–249.

[212] Nazareth, J.L. and Tseng, P. (2002), "Gilding the lily: a variant of the Nelder–Mead algorithm based on golden-section search," *Computational Optimization and Applications*, 22, 123–144.

[213] Nazareth, J.L. and Zhu, M. (1995), "Self-complementary variable metric algorithms," Technical Report 95-2. Department of Pure and Applied Mathematics, Washington State University, Pullman, Washington (revised, 1997).

[214] Nelder, J.A. and Mead, R. (1965), "A simplex method for function minimization," *Computer Journal*, 7, 308–313.

[215] Nemirovsky, A.S. and Yudin, D.B. (1983), *Problem Complexity and Method Efficiency in Optimization*, Wiley-Interscience, New York.

[216] Nesterov, Y.E. (1983), "A method of solving a convex programming problem with convergence rate $O(1/k^2)$," *Soviet Mathematics Doklady*, 27, 372–376.

[217] Nesterov, Y.E. and Nemirovsky, A.S. (1994), *Interior Point Polynomial Algorithms in Convex Programming*, SIAM Studies in Applied Mathematics, Vol. 13, SIAM, Philadelphia.

[218] Newell, A., Perlis, A.J., and Simon, H.A. (1967), "Computer Science," *Science*, 157, 1373–1374.

[219] Nielsen M.A. and Chuang, I.L. (2000), *Quantum Computation and Quantum Information*, Cambridge University Press, Cambridge, United Kingdom.

[220] O'Leary, D.P. (1996), "Conjugate gradients and related KMP algorithms: the beginnings," in *Linear and Nonlinear Conjugate Gradient-Related Methods*, L. Adams and J.L. Nazareth (eds.), SIAM, Philadelphia, 1–8.

[221] Oliveira, P.R. and Neto, J.X. (1995), "A unified view of primal methods through Riemannian metrics," Report ES-363/95, Program for Engineering Systems and Computation, Federal University of Rio de Janeiro, Brazil.

[222] Optimization Subroutine Library (1991), Guide and References, IBM Corporation, Kingston, New York.

[223] Oren, S.S. and Luenberger, D.C. (1974), "Self-scaling variable metric (SSVM) algorithms. Part 1: Criteria and sufficient conditions for scaling a class of algorithms, Part 2: Implementation and experiments," *Management Science*, 20, 845–862, 863–874.

[224] Ortega, J.M. and Rheinboldt, W.C. (1970), *Iterative Solution of Nonlinear Equations in Several Variables*, Academic Press, New York.

[225] Parlett, B.N. (1980), *The Symmetric Eigenvalue Problem*, Prentice-Hall, Englewood Cliffs, New Jersey.

[226] Penrose, R. (1989), *The Emperor's New Mind: Concerning Computers, Minds, and the Laws of Physics*, Oxford University Press, Oxford, England.

[227] Perry, A. (1977), "A class of conjugate gradient methods with a two-step variable metric memory," Discussion Paper 269, Center for Mathematical Studies in Economics and Management Science, Northwestern University, Evanston, Illinois.

[228] Perry, A. (1978), "A modified CG algorithm," *Operations Research*, 26, 1073–1078.

[229] Polak, E. (1971), *Computational Methods in Optimization: A Unified Approach*, Academic Press, New York.

[230] Polak, E. (1997), *Optimization: Algorithms and Consistent Approximations*, Series in Applied Mathematical Sciences, No. 124, Springer-Verlag, New York.

[231] Polak, E. and Ribière, G. (1969), "Note sur la convergence de méthode de directions conjuguées," *Revue Française d'Informatique et de Recherche Opérationnelle*, 16, 35–43.

[232] Polyak, B.T. (1964), "Some methods of speeding up the convergence of iteration methods," *USSR Comp. Math. and Math. Phys.*, 4, 1–17.

[233] Polyak, B.T. (1969), "The conjugate gradient method in extremal problems," *USSR Comp. Math. and Math. Phys.*, 9, 94–112.

[234] Polyak, B.T. (1987), *Introduction to Optimization*, Optimization Software Inc., Publications Division, New York.

[235] Polyak, R. (1992), "Modified shifted barrier function (theory and methods)," *Mathematical Programming*, 54, 177–222.

[236] Potra, F.A. (1996), "An infeasible interior point predictor–corrector algorithm for linear programming," *SIAM J. on Optimization*, 6, 19–32.

[237] Powell, M.J.D. (1964), "An efficient method for finding the minimum of a function of several variables without calculating derivatives," *Computer Journal*, 7, 155–162.

[238] Powell, M.J.D. (1969), "A method for nonlinear constraints in minimization problems," in *Optimization*, R. Fletcher (ed.), Academic Press, New York, 224–248.

[239] Powell, M.J.D. (1972), "Unconstrained minimization and extensions for constraints," A.E.R.E. Report TP 495, Harwell, England.

[240] Powell, M.J.D. (1977), "Restart procedures for the conjugate gradient method," *Mathematical Programming*, 12, 241–254.

[241] Powell, M.J.D. (1984), "Nonconvex minimization calculations and the conjugate gradient method," in: Lecture Notes in Mathematics, Vol. 1066, Springer-Verlag, Berlin, 122–141.

[242] Powell, M.J.D. (1986), "How bad are the BFGS and DFP methods when the objective function is quadratic?," *Mathematical Programming*, 34, 34–47.

[243] Powell, M.J.D. (1991), "A view of nonlinear optimization," in *History of Mathematical Programming: A Collection of Personal Reminiscences*, J.K. Lenstra, A.H.G. Rinnooy Kan, and A. Schrijver (eds.), North-Holland, Amsterdam, 119–125.

[244] Powell, M.J.D. (1998), "Direct search algorithms for optimization calculations," *Acta Numerica*, 7, 287–336.

[245] Powell, M.J.D. (2000), "UOBYQA: unconstrained optimization by quadratic approximation," Numerical Analysis Report DAMTP 2000/NA 14, Department of Applied Mathematics and Theoretical Physics, University of Cambridge, England.

[246] Press, W.H., Vetterling, W.T., Teukolsky, S.A., and Flannery, B.P. (1992), *Numerical Recipes in Fortran: The Art of Scientific Computing*, second edition, Cambridge University Press, Cambridge, England.

[247] Raveché, H.J., Lawrie, D.H., and Despain, A.M. (1987), *A National Computing Initiative*, Report of the Panel on Research Issues in Large-Scale Computational Science and Engineering, SIAM, Philadelphia.

[248] Rendl, F. and Wolkowicz, H. (1997), "A semidefinite framework for trust region subproblems with applications to large scale minimization," *Mathematical Programming*, 77, 273–300.

[249] Renegar, J. (1988), "A polynomial-time algorithm, based on Newton's method, for linear programming," *Mathematical Programming*, 40, 59–93.

[250] Renegar, J. (1995a), "Linear programming, complexity theory, and elementary functional analysis," *Mathematical Programming*, 70, 279–351.

[251] Renegar, J. (1995b), "Incorporating condition measures into the complexity theory of linear programming," *SIAM J. on Optimization*, 5, 506–524.

[252] Renegar, J. and Shub, M. (1992), "Unified complexity analysis for Newton LP methods," *Mathematical Programming*, 53, 1–16.

[253] Renegar, J., Shub, M., and Smale, S., eds. (1996), *Mathematics of Numerical Analysis*, Lecture Notes in Applied Mathematics, The American Mathematical Society, Providence, Rhode Island.

[254] Rice, J.R. (1971), "The challenge of mathematical software," in *Mathematical Software*, J. Rice (ed.), Academic Press, New York, 27–41.

[255] Rockafellar, R.T. (1976), "Augmented Lagrangians and applications of the proximal point algorithm in convex programming," *Mathematics of Operations Research* 1, 97–116.

[256] Rockafellar, R.T. (1993), "Lagrange multipliers and optimality," *SIAM Review*, 35, 183–238.

[257] Roos, C. and den Hertog, D. (1989), "A polynomial method of approximate weighted centers for linear programming," Report 99-13, Faculty of Technical Mathematics and Informatics, Delft University of Technology, Delft, the Netherlands.

[258] Rosenbrock, H.H (1960), "An automatic method for finding the greatest or least value of a function," *Computer Journal*, 3, 175–184.

[259] Saigal, R. (1992), "A simple proof of primal affine scaling method," Technical Report, Department of Industrial and Operations Engineering, University of Michigan, Ann Arbor, Michigan.

[260] Saigal, R. (1995), *Linear Programming*, Kluwer Academic Publishers, Dordrecht and Boston.

[261] Saunders, M.A. (1994), "Major Cholesky would feel proud," *ORSA J. on Computing*, 6, 23–27.

[262] Schnabel, R.B. (1994), "Parallel nonlinear optimization: limitations, opportunities and challenges," Report CU-CS-715-94, Department of Computer Science, University of Colorado, Boulder, Colorado.

[263] Schwefel, H.-P. (1981), *Numerical Optimization for Computer Models*, Wiley, Chichester, United Kingdom.

[264] Schwefel, H.-P. (1995), *Evolution and Optimum Seeking*, Wiley, Chichester, United Kingdom.

[265] Seifi, A. and Tuncel, L. (1996), "A constant-potential infeasible-start interior-point algorithm with computational experiments and applications," Research Report CORR 96-07, Department of Combinatorics and Optimization, University of Waterloo, Ontario, Canada.

[266] Shah, B.V., Buehler, R.J., and Kempthorne, O. (1964), "Some algorithms for minimizing a function of several variables," *J. Soc. Ind. and Appl. Math.*, 12, 74–91.

[267] Shanno, D.F. (1978), "Conjugate gradient methods with inexact searches," *Mathematics of Operations Research*, 3, 244–256.

[268] Shor, N.Z. (1985), *Minimization Methods for Non-Differentiable Functions*, Springer-Verlag, Berlin.

[269] Sikorski, K. (2001), *Optimal Solution of Nonlinear Equations*, Oxford University Press, Oxford, England.

[270] Smale, S. (1976), "A convergent process of price adjustment and global Newton methods," *Journal of Mathematical Economics*, 3, 1–14.

[271] Smith, B.T., Boyle, J.M., and Cody, W.J. (1974), "The NATS approach to quality software," in *Proceedings of IMA Conference on Software for Numerical Mathematics*, J. Evans (ed.), Academic Press, New York, 393–405.

[272] Sonnevend, G. (1986), "An 'analytic center' for polyhedrons and new classes of global algorithms for linear (smooth, convex) programming," in *System Modelling and Optimization*, A. Prekopa, J. Szelezsan, and B. Strazicky (eds.), Lecture Notes in Control and Information Sciences, Vol. 84, Springer-Verlag, Heidelberg and Berlin, 866–875.

[273] Sonnevend, G., Stoer, J., and Zhao, G. (1990), "On the complexity of following the central path of linear programs by linear extrapolation," *Methods of Operations Research*, 62, 19–31.

[274] Sonnevend, G., Stoer, J., and Zhao, G. (1991), "On the complexity of following the central path of linear programs by linear extrapolation II," *Mathematical Programming*, 52, 527–553.

[275] Spendley, W., Hext, G.R., and Himsworth, F.R. (1962), "Sequential application of simplex designs in optimisation and evolutionary operation," *Technometrica*, 4, 441–461.

[276] Stewart, I. (1987), *The Problems of Mathematics*, Oxford University Press, Oxford and New York.

[277] Strang, G. (1986), *Introduction to Applied Mathematics*, Wellesley-Cambridge Press, Cambridge, Massachusetts.

[278] Tanabe, K. (1988), "Centered Newton method for mathematical programming," in *Systems Modelling and Optimization*, M. Iri and K. Yajima (eds.), Lecture Notes in Control and Information Sciences, Vol. 113, Springer-Verlag, Heidelberg and Berlin, 197–206.

[279] Todd, M.J. (1996), "Potential-reduction methods in mathematical programming," *Mathematical Programming*, 76, 3–45.

[280] Todd, M.J. and Ye, Y. (1990), "A centered projective algorithm for linear programming," *Mathematics of Operations Research*, 15, 508–529.

[281] Torczon, V. (1989), *Multi-Directional Search: A Direct Search Algorithm for Parallel Machines*, Ph.D. dissertation, Rice University, Houston, Texas; available as Technical Report 90-7, Department of Mathematical Sciences, Rice University, Houston, Texas.

[282] Traub, J.F. (1964), *Iterative Methods for the Solution of Equations*, Prentice-Hall, Englewood Cliffs, New Jersey.

[283] Traub, J.F. and Werschulz, A.G. (1998), *Complexity and Information*, Cambridge University Press, Cambridge, England.

[284] Trefethen, L.N. (1992), "The definition of numerical analysis," *SIAM News*, 25, p. 6.

[285] Tseng, P. (1992), "Complexity analysis of a linear complementarity algorithm based on a Lyapunov function," *Mathematical Programming*, 53, 297–306.

[286] Tseng, P. (1995), "Simplified analysis of an $O(nL)$-iteration infeasible predictor–corrector path-following method for monotone linear complementarity problems," in *Recent Trends in Optimization Theory and Applications*, R.P. Agarwal (ed.), World Scientific Publishing Company, 423–434.

[287] Tseng, P. (2000), "Fortified-descent simplicial search method: a general approach," *SIAM J. on Optimization*, 10, 269–288.

[288] Tsuchiya, T. and Muramatsu, M. (1995), "Global convergence of a long-step affine scaling algorithm for degenerate linear programming problems," *SIAM J. on Optimization*, 5, 525–551.

[289] Tuncel, L. (1994), "Constant potential primal–dual algorithms: a framework," *Mathematical Programming*, 66, 145–159.

[290] Vanderbei, R.J. (1994), "Interior point methods: algorithms and formulations," *ORSA J. on Computing*, 6, 32–43.

[291] Vanderbei, R.J. (1995), "LOQO, an interior point code for quadratic programming," Technical Report, Program in Statistics and Operations Research, Princeton University, Princeton, New Jersey.

[292] Vanderbei, R.J. (1996), *Linear Programming*, Kluwer Academic Publishers, Dordrecht and Boston.

[293] Vanderbei, R.J., Meketon, M.S., and Freedman, B.A. (1986), "A modification of Karmarkar's linear programming algorithm," *Algorithmica*, 1, 395–407.

[294] Wang, W. and O'Leary, D.P. (1995), "Adaptive use of iterative methods in interior point methods for linear programming," Applied Mathematics Program, University of Maryland, College Park, Maryland.

[295] Watson, L.T. (1986), "Numerical linear algebra aspects of globally convergent homotopy methods," *SIAM Review*, 28, 529–545.

[296] Watson, L.T., Billups, S.C., and Morgan, A.P. (1987), "HOMPACK: A suite of codes for globally convergent homotopy algorithms," *ACM Transactions on Mathematical Software*, 13, 281–310.

[297] Weerawarana, S., Houstis, E.N., Rice, J.R., Joshi, A., and Houstis, C.E. (1996), "PYTHIA: A knowledge-based system to select scientific algorithms," *ACM Transactions on Mathematical Software*, 22, 447–468.

[298] Wilkinson, J.H. (1963), *Rounding Errors in Algebraic Processes*, Prentice-Hall, Englewood Cliffs, New Jersey.

[299] Wilkinson, J.H. (1965), *The Algebraic Eigenvalue Problem*, Clarendon Press, Oxford, England.

[300] Wolfe, P. (1969), "Convergence conditions for ascent methods," *SIAM Review*, 11, 226–235.

[301] Xu, X., Hung, P-F., and Ye, Y. (1996), "A simplification of the homogeneous and self-dual linear programming algorithm and its implementation," *Annals of Operations Research*, 62, 151–172.

[302] Ye, Y. (1991), "An $O(n^3L)$ potential reduction algorithm for linear programming," *Mathematical Programming*, 50, 239–258.

[303] Ye, Y., Todd, M.J., and Mizuno, S. (1994), "An $O(\sqrt{n}L)$-iteration homogeneous and self-dual linear programming algorithm," *Mathematics of Operations Research*, 19, 53–67.

[304] Yuan, Y. and Byrd, R.H. (1995), "Non-quasi-Newton updates for unconstrained optimization," *Journal of Computational Mathematics*, 13, 95–107.

[305] Zangwill, W.I. (1967), "Minimizing a function without calculating derivatives," *Computer Journal*, 10, 293–296.

[306] Zangwill, W.I. (1969), *Nonlinear Programming: A Unified Approach*, Prentice Hall, Englewood Cliffs, New Jersey.

[307] Zhang, Y. (1994), "On the convergence of a class of infeasible interior-point algorithms for the horizontal complementarity problem," *SIAM J. on Optimization*, 4, 208–227.

[308] Zhang, Y. (1995), "LIPSOL: a MATLAB toolkit for linear programming," Department of Mathematics and Statistics, University of Maryland, Baltimore County, Maryland.

[309] Zhang, Y. and Zhang, D. (1995), "On polynomiality of the Mehrotra-type predictor–corrector interior-point algorithms," *Mathematical Programming*, 68, 303–318.

[310] Zhu, M. (1997), *Limited memory BFGS Algorithms with Diagonal Updating*, M.Sc. Project in Computer Science, School of Electrical Engineering and Computer Science, Washington State University, Pullman, Washington.

[311] Zhu, M. (1997), *Techniques for Nonlinear Optimization: Principles and Practice*, Ph.D. dissertation, Department of Pure and Applied Mathematics, Washington State University, Pullman, Washington.

[312] Zhu, M., Nazareth, J.L., and Wolkowicz, H. (1999), "The quasi-Cauchy relation and diagonal updating," *SIAM J. on Optimization*, 9, 1192–1204.

Index

MIX
Papier aus verantwortungsvollen Quellen
Paper from responsible sources
FSC® C105338

If you have any concerns about our products,
you can contact us on
ProductSafety@springernature.com

In case Publisher is established outside the EU,
the EU authorized representative is:
Springer Nature Customer Service Center GmbH
Europaplatz 3, 69115 Heidelberg, Germany

Printed by Libri Plureos GmbH
in Hamburg, Germany